Nuclear Non-Proliferation

M. P. Fry N. P. Keatinge J. Rotblat (Eds.)

Nuclear Non-Proliferation

and the Non-Proliferation Treaty

With 7 Figures

Springer-Verlag Berlin Heidelberg New York
London Paris Tokyo Hong Kong

Dr. Michael P. Fry
Department of Theoretical Physics, University of Dublin, 39 Trinity College, Dublin 2, Ireland

Dr. Patrick Keatinge
Department of Political Science, University of Dublin, 39 Trinity College, Dublin 2, Ireland

Professor Dr. Joseph Rotblat
Pugwash Conferences on Science and World Affairs, Flat A, Museum Mansions, 63A Great Russell Street, London, WC1B 3BJ, United Kingdom

ISBN-13: 978-3-642-75107-3 e-ISBN-13: 978-3-642-75105-9
DOI: 10.1007/978-3-642-75105-9

Library of Congress Cataloging-in-Publication Data. Nuclear non-proliferation and the non-proliferation treaty / edited by M. P. Fry, N. P. Keatinge, and J. Rotblat. p. cm. Includes bibliographical references. ISBN 0-387-51756-1 (U.S. : alk. paper) 1. Nuclear nonproliferation. I. Fry, M. P. (Michael P.), 1940– . II. Keatinge, Patrick, 1939– . III. Rotblat, Joseph, 1908– . JX1974.73.N775 1990 327.1'74–dc20 89-26202

Softcover reprint of the hardcover 1st edition 1990

2157/3150(3011)-543210 – Printed on acid-free paper

Foreword

This volume appears at a time when the prospects for banishing the threat of nuclear annihilation are brighter than at any time since the first atomic device exploded over the desert at Alamogordo. The last few years have seen an extraordinary change in the climate of East-West relations. The programme of political and economic reform which President Gorbachev initiated in the Soviet Union and which is now spreading throughout most of Eastern Europe has been parallelled by serious efforts to reach agreement on measures for conventional and nuclear disarmament. This has led to new hope that international peace and security can at last be built upon the firm foundation of justice, respect for international law and a determination to approach problems in a spirit of genuine co-operation rather than one of distrust and confrontation.

This new climate encourages us in the belief that the obvious common sense of preventing the further spread of nuclear weapons will come to be shared by all nations.

At the same time, we have to recognize two very disturbing facts, which imply that there can be no slackening of our efforts to strengthen the non-proliferation regime.

First, despite the great improvement in East-West relations and despite the recognition by former President Reagan and President Gorbachev that a nuclear war cannot be won and must never be fought, the reality is that the nuclear arms race has not yet come to an end. The reports of progress at the START negotiations are, of course, most welcome and encouraging. Nevertheless, even a successful conclusion to these negotiations and a 50% reduction in strategic nuclear weapons would still leave a world with nuclear arsenals as great as they were when the first Strategic Arms Limitation Treaty was signed in 1972. The sad fact is that nations still rely on doctrines of deterrence based on their own military strength, and have not yet turned decisively towards a policy which sees the elimination of the means of waging offensive war as the only certain shield against aggression.

Second, there is a very real danger that further states will take the fateful step of acquiring nuclear weapons. It is true that no country beyond the five existing nuclear powers has declared itself to possess a nuclear weapon capacity. Nevertheless, there are persistent and deeply worrying reports that a certain number of states have clandestinely, or in some cases with only feeble efforts to cloak their activities, moved very close indeed to the nuclear threshold.

In our view, the conclusions to be drawn from these two disturbing facts are very clear. As long as nuclear weapons continue to exist there is the danger that one day they may be used. As long as some states continue to possess nuclear weapons, others will be tempted to follow in the same path. As I said in my address to the UN Third Special Session on Disarmament in June 1988, the only acceptable level of nuclear weapons is zero.

What can be done to bring about this desired goal?

First, I think it has to be recognized that the existing nuclear powers bear the greatest responsibility in pushing back the menace of the further growth and spread of nuclear arms, whether among themselves or in other parts of the world. Just as those powers gave the lead in introducing these weapons of mass destruction into their arsenals, so must they lead in halting and reversing this arms race, which holds such terrible potential for the destruction of the world we all share.

We find it particularly disappointing that, after so many years of efforts, the promise to achieve a Comprehensive Nuclear Test Ban Treaty has not yet been fulfilled. Instead, the number of test explosions carried out since the signing of the Limited Test Ban Treaty in 1963 has doubled, and the nuclear arsenals have grown from some hundreds of devices to over 50,000. The sole result has been a world of diminished security at higher levels of armaments.

The claim is sometimes advanced that nuclear weapons have helped to prevent the tensions of our world breaking out into open conflict. But surely it ought to be clear that one of the major causes of the persisting tension and threat of war is the arms race itself.

Despite our concern at the continuing disproportion between the achievements of the disarmament process so far and the alarming progress of the arms race, there can be no excuse for other countries to seek a nuclear weapon capacity of their own. The non-proliferation of nuclear weapons has become the accepted norm of international life, and it is both the responsibility and the interest of every country, large or small, nuclear or non-nuclear, to do all in its power to ensure that the menace of nuclear weapons will spread no further. Any breach of the non-proliferation regime both weakens the moral prohibition against nuclear proliferation and creates a direct security incentive for other states to pursue a similar policy, thus negating the supposed advantage which the original proliferator had hoped to gain.

It is also deeply to be regretted that a number of states have not yet adhered to the Nuclear Non-Proliferation Treaty. This Treaty, which resulted from an Irish initiative at the United Nations, has made an indispensable contribution to the cause of international peace and security. It is of particular importance that the non-proliferation regime should be strengthened. There are genuine security problems between countries, but these problems will only be deepened by the acquisition of nuclear weapons. They can never be resolved by military threat.

It is often said that these are very difficult and complex questions. In a technical sense, so they are. But the political principles are very clear, and the issue affects the lives of each and every one of us. These are not questions to be left to the statesmen or the generals or experts of any kind. The active political involvement of the public is essential.

The Pugwash Conferences have made significant contributions over the years to advancing public awareness and understanding. The holding of a symposium on Nuclear Non-Proliferation in Dublin in May 1989 was an important event, especially as we approach the Fourth NPT Review Conference in August of next year.

Ultimately, it is a question of choosing what kind of world we want to live in – a world dominated by confrontation and fear or a world of community and cooperation.

Dublin, December 1989

Charles Haughey
Prime Minister of Ireland

Preface

This volume comprises a selection of papers contributed to a Pugwash Symposium on Nuclear Non-Proliferation and the Non-Proliferation Treaty, held in Dublin, Ireland, from 5 to 7 May 1989. Its purpose was to assess the future of the entire nuclear non-proliferation regime and to recommend guidelines for strengthening it in time for the 1990 Review Conference. The forty-five participants in the symposium were drawn from eighteen countries, including declared nuclear-weapon states and nuclear-weapon threshold states. The authors of the papers are in authority and/or authoritative on the subject of nuclear proliferation, by virtue of current or previous high office, or other special qualifications. A considerable amount of the material presented is new on an issue which, we believe, will increasingly dominate international relations in the 1990s. The following synopsis of the non-proliferation regime is meant to set the stage for the papers that follow. In preparing this outline the editors are indebted to Professor John Holdren who served as rapporteur and on whose statement we have drawn.

The 1968 Treaty on the Non-Proliferation of Nuclear Weapons (NPT, see Appendix) is the cornerstone of an international set of arrangements – the nuclear non-proliferation regime – to prevent the "horizontal" spread of nuclear weapons to other countries beyond the five declared nuclear-weapon states. Each nuclear-weapon state party to the NPT undertakes not to transfer or to give control of, directly or indirectly, nuclear weapons to any non-nuclear-weapon state, nor to assist or encourage in any way their manufacture by such a state (Article I). It is understood that this prohibition applies also to non-nuclear-weapon states. The Preamble of the Treaty states, and Article VI reaffirms, the intention of the nuclear-weapon states to achieve the cessation of the nuclear arms race at the earliest possible date and to undertake effective nuclear disarmament measures. As specific measures, the Preamble endorses a comprehensive ban on the testing of nuclear weapons (CTB) and their elimination from national arsenals, as well as their means of delivery. In turn, each non-nuclear-weapon state party to the NPT undertakes to forgo the acquisition of nuclear weapons by any means whatsoever (Article II) and to accept comprehensive safeguards of the International Atomic Energy Agency (IAEA) to prevent the diversion of nuclear energy from peaceful uses to nuclear weapons (Article III). As stated in the Preamble and reaffirmed in Articles IV and V, all the parties to the Treaty are entitled to develop and use nuclear energy for peaceful purposes, and nothing in the Treaty should infringe on this right.

The NPT entered into force in 1970 and has been signed and ratified by 140 states as of April 1989. Among the parties to the Treaty are three nuclear-weapon states – the United Kingdom, the United States and the Soviet Union – and the bulk of the highly industrialized, militarily significant states capable of producing nuclear weapons. France, a non-party, has stated that it would behave as if it were one, and China, also a non-party, has given solemn assurances that it will abide by non-proliferation principles. Article VIII of the Treaty provides for a review of its operation at intervals of five years, with the Fourth Review Conference scheduled for 1990. Article X states that a conference shall be called in 1995 "to decide whether the Treaty shall continue in force indefinitely, or shall be extended for an additional fixed period or periods". In theory, although it is not likely, the Treaty could be terminated by a majority of the parties.

There are other treaties dealing with nuclear proliferation in a direct way. These are the 1963 treaty banning Nuclear Weapons Tests in the Atmosphere, in Outer Space and Under Water (Limited Test Ban Treaty), the 1967 Treaty for the Prohibition of Nuclear Weapons in Latin America (Treaty of Tlatelolco) together with its Organization for the Prohibition of Nuclear Weapons in Latin America, and the 1986 South Pacific Nuclear Free Zone Treaty (Treaty of Rarotonga). Supplementing these are treaties and settlements, dating from World War II, restricting production and possession of nuclear weapons, as well as unilateral undertakings by a number of threshold states not to assist other states to acquire nuclear weapons.

The NPT has spurred the introduction of nuclear export controls. Article III.2 provides for safeguards on the export, for peaceful purposes, to any non-nuclear state of "source and special fissionable material" (defined in Article XX of the IAEA's Statute), and of sensitive equipment and components used in the processing and production of fissionable material. Clarification of the scope of Article III.2 resulted in the 1977 Guidelines for Nuclear Transfers, which are administered by state members of the Nuclear Supplier Group, the so-called London Club. Other countries have supplemented these guidelines by enacting their own, stricter export controls, such as the 1978 US Nuclear Non-Proliferation Act.

In addition to the IAEA's safeguards there are those of the European Atomic Energy Community (EURATOM). These apply to all nuclear material in the member states of the Community and operate in conjunction with the IAEA's safeguards through an agreement that coordinates both safeguards systems.

These arrangements constitute the bulk of the present international non-proliferation regime. As noted, it is not entirely accepted by all states. Argentina, Brazil, India, Israel, Pakistan and South Africa remain non-parties to the NPT. Each has, or is reported to have under construction, unsafeguarded facilities capable of producing weapon-usable material. Argentina and Pakistan have signed but not ratified the Limited Test Ban Treaty. Argentina has signed but not ratified the Treaty of Tlatelolco, while Brazil has ratified it but prevented its entering into force. France has signed but not ratified Protocol II of the Treaty. The Protocols of the Treaty of Rarotonga have not yet been signed by the USA, UK and

France. Finally, several nuclear supplier countries, including China, have not yet announced acceptance of the Guidelines for Nuclear Transfers.

Notwithstanding the above holdbacks, the non-proliferation regime is an impressive achievement. As pointed out in Mr. Ove Bring's paper, the non-proliferation regime is more than a treaty regime in the sense that, with the passage of time, it is becoming rooted in customary international law, thereby acquiring a life of its own. This, and the perception by particular states that the acquisition of nuclear weapons would lessen their security, is a cause for cautious optimism.

Part One of this volume surveys recent technical, economic and legal factors bearing on the non-proliferation regime, including the NPT and nuclear trade. Plutonium recycling and emerging technologies, such as laser enrichment facilities, in addition to the continued worldwide growth of the nuclear power industry, are posing an increasing challenge to the IAEA's safeguards system. Appropriately, a discussion of the adequacy of the IAEA's safeguards for the 1990s is included here.

Part Two is primarily concerned with the nuclear-weapon threshold states and their incentives and disincentives for entering into a framework of restraint. By definition, the non-proliferation regime is larger than the NPT; there remains substantial room for new initiatives to reinforce the regime. These include: regional measures, such as nuclear-weapon-free zones; declarations by neighbouring countries to forgo the acquisition of nuclear weapons and to open their nuclear facilities to mutual inspection; greater willingness on the part of the industrialized countries to help less developed countries develop their nuclear power industries under appropriate safeguards; and security guarantees from nuclear-weapon states not to use nuclear weapons against non-nuclear-weapon states that renounce these weapons. The feasibility of these and other non-proliferation measures are examined here.

Part Three recognizes that the non-proliferation regime is an integral part of the prevailing international order whose parameters are rapidly changing. Central to this order is the most rapidly changing parameter of all, the USA-USSR relationship. Both states have found it in their mutual interest to support the goals of non-proliferation. The development of their active cooperation on this matter is dealt with in one of the papers in this section. The other two consider the likely impact of improved Soviet-American relations on nuclear proliferation in Central Europe, the Middle East, the Pacific Basin and in Africa.

The NPT is asymmetric in the rights and obligations it defines for nuclear-weapon and non-nuclear-weapon states party to it. Its manifest discrimination between the two is the basis of deep reservations about the Treaty by non-parties, and is the source of increasing dissatisfaction by many of its non-weapon adherents. As the Preamble and Article VI affirm, the distinction between weapon and non-weapon parties is supposed to be temporary while the process of nuclear disarmament proceeds. This distinction has hardly changed after nineteen years. Some would even argue that the Intermediate Nuclear Forces (INF) Treaty represents no progress at all, since it abolishes a category of nuclear weapons that did not exist when the NPT entered into force in 1970. In brief, nuclear-weapon states party to the NPT are charged with not fulfilling their obligations while holding the non-weapon parties to theirs.

Part Four discusses the priorities for pursuing the disarmament obligations set forth in the NPT's Preamble and Article VI. Inevitably through it all runs the scarlet thread of a comprehensive ban on nuclear testing. Whether the non-weapon parties to the NPT would consider a CTB as substantive progress towards nuclear disarmament is a moot point. Certainly a CTB imposing identical obligations on weapon and non-weapon parties could not be labeled as "discriminatory". A nearly universally adhered to and verifiable CTB would act as a strong brake on any state seeking to conduct a nuclear test, and would be an important addition to the non-proliferation regime. Verification of a CTB and a possible means of achieving such a treaty, by converting the 1963 Limited Test Ban Treaty into a CTB by amendment, are accordingly included in this section.

Viewed in perspective, the admirable goal of preventing the "horizontal" spread of nuclear weapons is a narrow one. Granted that there is no logical basis for denying the "right" of the nuclear deterrent to some while according it to others, the only truly stable non-proliferation regime is a world without nuclear weapons. How this might be realistically achieved is the subject of the remaining papers in this section. Attention naturally focuses on the Indian Proposal at the Third United Nations Special Session on Disarmament, calling for the 1995 NPT Conference to replace the Treaty with a new one creating a comprehensive system of international peace and security in which all states participate on an equal footing, and which would lead to the elimination of nuclear weapons by 2010.

Dublin and London, October 1989 *Fry, Keatinge, Rotblat*

Contents

Part III Nuclear Proliferation and Soviet-American Relations

Part IV Priorities for Pursuing the Objectives of Article VI of the NPT

Berger, Jonathan

Geophysics, Institute of Geophysics, Scripps Institution of Oceanography, University of California, San Diego, La Jolla, CA 92093, USA

Bring, Ove

International Law, Legal Adviser, Ministry for Foreign Affairs, Gustav Adolfs torg 1, Box 16121, S-10323 Stockholm, Sweden

Carasales, Julio

Ambassador, Ministry of Foreign Affairs, Buenos Aires, Argentina

Davydov, Valerii

Arms Control, Institute of the USA and Canada Studies, USSR Academy of Sciences, Khlebny per 2/3, 121069 Moscow, USSR

Epstein, William

Arms Control, UN Institute for Training and Research, 801 UN Plaza, New York, NY 10017, USA; former Head of Disarmament Affairs, UN

Fischer, David

Economics, History; 15 Willow Walk, Cambridge CB1 1LA, United Kingdom; former Assistant Director-General, International Atomic Energy Agency, Austria

Freier, Shalheveth

Physics, Weizmann Institute of Science, Rehovot 76100, Israel; former Head of Israel Atomic Energy Commission

Fry, Michael

Theoretical Physics, University of Dublin, 39 Trinity College, Dublin 2, Ireland

Goldanskii, Vitalii

Physics and Chemistry, Director, N. N. Semenov Institute of Chemical Physics, USSR Academy of Sciences, Vlitsa Kosygina, 4; 117334 Moscow, USSR

Goldblat, Jozef

Arms Control, Graduate Institute of International Studies, 132, rue de Lausanne, 1202 Geneva, Switzerland

Keatinge, Patrick

Political Science, University of Dublin, Trinity College, Dublin 2, Ireland

Khan, Munir Ahmad

Nuclear Energy, Chairman, Pakistan Atomic Energy Commission, P.O. Box No. 1114, Islamabad, Pakistan

Mahfouz, Mahmoud

Medicine, Radiation Oncology; 15 Sherif Street, Cairo, Egypt; former Minister of Health

Merlini, Cesare

Engineering, President, Institute of International Affairs, Viale Mazzini, 88; 00195 Rome, Italy

Müller, Harald

Political Science, Director of International Programmes, Peace Research Institute, Leimenrode 29, D-6000 Frankfurt/Main 1, Fed. Rep. of Germany

Prawitz, Jan

Arms Control, Adviser to Ministry of Defence, S-10333 Stockholm, Sweden

Rotblat, Joseph

Physics, Emeritus Professor of Physics, University of London; President, Pugwash Conferences on Science and World Affairs, Flat A, Museum Mansions, 63A Great Russell Street, London WC1B 3JB, United Kingdom

Sanders, Ben

International Relations, Director, Programme for Promoting Nuclear Non Proliferation, 240 East 27th Street, USA; former Director, Information and Studies, Department of Disarmament Affairs, UN, New York, USA

Taylor, Theodore

Physics, Independent Consultant, P.O. Box 39, West Clarkesville, NY 14786, USA; former Deputy Director, Defence Nuclear Agency

Udgaonkar, Bhalchandra

Theoretical Physics, Senior Professor, Tata Institute of Fundamental Research, Homi Bhabba Road, Bombay 400005, India

Wilmshurst, Michael

International Relations, Director, External Relations, International Atomic Energy Agency, Wagramerstrasse 5, P.O. Box 100, A-1400 Vienna, Austria

Acronyms

ACDA	Arms Control and Disarmament Agency
ANS	American Nuclear Society
ANZUS	Australia, New Zealand and the United States (Signatories to the Tripartite Security Treaty of 1951)
C^3I	command, control, communications, and intelligence
CCD	Conference of the Committee on Disarmament
CD	Committee on Disarmament
CFE	Conventional Forces in Europe
CTB	comprehensive test ban
ENDC	Eighteen-Nation Committee on Disarmament
HEU	highly enriched uranium
HWR	heavy-water reactor
IAEA	International Atomic Energy Agency
IBRD	International Bank for Reconstruction and Development
IISS	International Institute for Strategic Studies
INF	intermediate nuclear forces
LDC	less developed country
LWR	light-water reactor
MW(e)	one million (electric) watts
NNA	neutral and non-aligned
NNWS	non-nuclear weapon state(s)
NPT	Non-Proliferation Treaty
NWFZ	nuclear-weapon-free zone
NWS	nuclear weapon state(s)
OECD	Organisation for Economic Cooperation and Development
OPANAL	Organization for the Prohibition of Nuclear Weapons in Latin America
PNE	peaceful nuclear explosion
PTBT	Partial Test Ban Treaty
PWR	pressurized-water reactor
SDI	Strategic Defense Initiative
SIPRI	Stockholm International Peace Research Institute
SNF	short-range nuclear forces
SWU	separative work units
TOE	tonnes of oil equivalent
UKAEA	United Kingdom Atomic Energy Authority

UNCTAD	United Nations Conference on Trade and Development
UNSSOD	United Nations Special Session on Disarmament
VVER	a standard design pressurized-water reactor manufactured in the USSR

Part I

Nuclear Proliferation: Technical and Economic Aspects

The Non-Proliferation Treaty: Status of Implementation and the Threatening Developments

Jozef Goldblat

The need to restrain the military threat of nuclear energy has been evident to many people from the early days of the atomic age. Indeed, the very first UN General Assembly resolution, of January 1946, called for the elimination of nuclear weapons from state arsenals. In the same year, the government of the United States, which was the first to manufacture these weapons and to use them, proposed the establishment of an international authority to control all atomic energy activities. This proposal, known as the Baruch Plan, met with no success. In 1949 the Soviet Union also became a nuclear weapon power, followed in 1952 by the United Kingdom, in 1960 by France, and in 1964 by China. Subsequently, however, the realization that proliferation of nuclear weapons would pose a danger to world security led to the development of a non-proliferation regime which encompasses various restrictive rules as well as specialized control institutions, both national and international. Among the latter, the International Atomic Energy Agency (IAEA) fulfils an essential practical role, but the pivotal place in the regime belongs to the Treaty on the Non-Proliferation of Nuclear Weapons (NPT), signed in 1968. The NPT is a unique document in the sense that it prohibits the possession by an overwhelming majority of states of the most destructive weapons yet invented, while tolerating the retention of the same weapons by a handful of nations. But the NPT is not an end in itself: the declared aim of the parties is to use it as a transitional measure to clear the way towards nuclear disarmament.

In spite of the inequality of treaty rights and obligations of the nuclear and non-nuclear weapon parties, the NPT, in force since 1970, has attracted a record number of adherents for an arms control agreement – nearly 140. These include three nuclear weapon powers – the United Kingdom, the United States and the Soviet Union – as well as almost all highly developed, industrialized and militarily significant non-nuclear weapon states. France, a nuclear weapon power which has not signed the NPT, has a declared policy of behaving like a state party to it. China, the fifth nuclear weapon power, has given solemn assurances that it would not help other states to acquire nuclear weapons.

In the course of the past two decades the non-proliferation regime has been considerably strengthened. No material breaches of the NPT have been recorded, and no withdrawals have been notified. However, the non-proliferation regime is also experiencing a few disquieting trends. This is due to the planned acquisition of nuclear-powered submarines by non-nuclear-weapon states, the growing trade in nuclear-capable missiles and the emergence of new suppliers of nuclear hardware and services.

Status of Implementation

Non-Transfer and Non-Acquisition of Nuclear Weapons. The essential non-proliferation undertakings are contained in the first two articles of the NPT. Article I places the nuclear-weapon states under the obligation not to transfer "to any recipient whatsoever" nuclear weapons or other nuclear explosive devices or control over them, and not in any way to "assist, encourage, or induce" any non-nuclear weapon state to manufacture or acquire such weapons or devices. Article II pledges the non-nuclear-weapon states not to receive nuclear weapons or other nuclear explosive devices or control over them, as well as not to manufacture them or receive assistance in their manufacture.

The implementation of the first part of Article I prohibiting transfer of nuclear explosive devices, though unverifiable, has not given rise to formal complaints. The extent to which the second part of Article I has been observed – that part prohibiting the provision by the nuclear-weapon states of assistance in the manufacture of nuclear weapons – has lent itself to controversy. Since there exists a significant overlap between the technologies of civilian nuclear energy and those useful for military explosive purposes, and since nuclear material and technology nominally destined for power programmes have been exported by NPT parties to countries which have not formally forgone nuclear weapons, it is argued by some that the obligation not in "any way" to assist non-nuclear-weapon states to manufacture nuclear explosive devices has not been fully complied with.

As regards Article II, there is no evidence that any non-nuclear-weapon party to the NPT has clandestinely manufactured or otherwise acquired nuclear explosive devices. If any one of them has designed a nuclear weapon or even developed its non-nuclear components, these activities would be difficult to detect. Should such a state ever decide actually to produce a nuclear weapon, it would need the requisite quantity of weapon-grade fissile material. The availability of this material is, therefore, of crucial significance; hence the importance of safeguards to prevent its diversion from peaceful to military uses.

Nuclear Safeguards and Protection of Nuclear Material. The safeguards requirement under Article III constitutes the verification element of the NPT. Safeguards should enable detection of diversion of significant quantities of nuclear material from peaceful activities to the manufacture of nuclear explosive devices, as well as deterrence of diversion by creating the risk of timely detection. No such diversion has as yet been reported by the IAEA, although on several occasions the Agency has been hindered in its inspection activities.

The Treaty requires safeguards to be implemented in such a manner as to avoid hampering the economic or technological development of the countries party to it, or international co-operation in the field of peaceful nuclear activities. This requirement seems to have been met, although there have been some complaints that controls complicate the production process, or are a burden for enterprises in terms of cost and of protection of industrial secrets. More controversial is the clause setting forth the conditions for nuclear trade with non-nuclear weapon countries. This clause has been applied in a way that has sometimes benefited non-parties more than parties. For whereas parties are subject to NPT safeguard covering all their peaceful nuclear activities, the nuclear activities of non-parties are covered only partially, by safeguards of the pre-NPT order, which apply exclusively to imported items – individual installations or material – while part of the nuclear fuel cycle may remain unsafeguarded. Many suppliers concerned about the dangers of nuclear proliferation inherent in the distinction between imported and domestic technology have sought to impose on non-parties full-scope safeguards, as extensive as NPT-type safeguards. A few suppliers, however, are reluctant to modify radically their export conditions.

An important step towards reducing the risks of diversion of nuclear material to non-peaceful purposes was made in 1987, with the entry into force of the 1980 Convention on the Physical Protection of Nuclear Material. The provisions of the Convention oblige the parties to ensure that, during international transport across their territory or on ships or aircraft under their jurisdiction, nuclear material for peaceful purposes, as categorized in a special annex (plutonium, uranium-235, uranium-233 and irradiated fuel), is protected at the agreed level. Furthermore, the parties undertake not to export or import nuclear material or allow its transit through their territory unless they have received assurances that this material will be protected during international transport in accordance with the levels of protection determined by the Convention. The parties to the Convention agree to share information on missing nuclear material to facilitate recovery operations. Robbery, embezzlement or extortion in relation to nuclear material, and acts without lawful authority involving nuclear material which cause or are likely to cause death or serious injury to any person or substantial damage to property, are to be treated as punishable offences.

Towards the end of 1987, alarm was raised in the FRG and Belgium because of alleged illegal cross-border transportation of canisters with nuclear wastes. It was asserted that at least some canisters were falsely labelled and actually contained fissionable material destined for Pakistan and Libya. No evidence was supplied to support this allegation [1]. It seems, none the less, that some serious irregularities in the transportation of radioactive substances did take place. The members of the European community have not yet ratified the Physical Protection Convention, even though shipment of nuclear material in the territories of the Community is very intensive.

Peaceful Uses of Nuclear Energy. Article IV of the NPT reaffirms the right of parties to develop nuclear energy for peaceful purposes in conformity with Articles I and II of the Treaty and obligates those parties in a position to do so to contribute to such efforts in non-nuclear-weapon states. The implementation of

Article IV has been affected to a great extent by the world-wide slow-down in the growth of civilian nuclear power owing to environmental, safety and economic factors, which include a weak increase in electricity demand, high initial investment costs and shortage of capital. In addition, restrictions have been imposed on nuclear supplies ever since India, a non-party to the NPT, took advantage of the lenient pre-NPT safeguards applied to its nuclear activities to explode a nuclear device. In 1977, a group of nuclear suppliers, the so-called London Club, drew up a list of materials, equipment and technology which should "trigger" IAEA safeguards when exported to non-nuclear-weapon states not party to the NPT (the parties having already accepted the requirement for safeguards on all their nuclear activities). The Guidelines for Nuclear Transfers agreed by the London Club require the recipients of the trigger-list items to provide effective physical protection for these items, and to pledge not to use them for the manufacture of nuclear explosives. The safeguards requirements apply to any "replicated" facility, that is, of the same type as the imported facility but constructed indigenously within a specified period. Retransfers of trigger-list items are to be subject to the same conditions as those attached to the original transfer. In the event of the diversion of materials or a violation of the supplier/recipient understandings, the members of the London Club should consult promptly on possible common action. Moreover, "restraint" is recommended in the transfer of sensitive facilities, such as uranium enrichment and plutonium reprocessing plants. Significantly, this restraint is to be exercised not only with respect to non-parties, but also with respect to parties to the NPT. In 1978 the USA unilaterally set even sterner restrictions on nuclear supplies by adopting the Nuclear Non-Proliferation Act (NNPA).

Spokesmen of many countries, especially from the Third World, have criticized the restrictive measures taken by the suppliers as an infringement of the right to nuclear supplies implied in NPT Article IV. Their argument is that, once governments have accepted the safeguards provided for in Article III, no further limitation should be placed on peaceful nuclear programmes. The Committee on Assurances of Supply which was set up by the IAEA to consider and advise on "ways and means in which supplies of nuclear material, equipment and technology and fuel cycle services could be assured on a more predictable and long-term basis in accordance with mutually acceptable considerations of non-proliferation" has not, as yet, produced agreed principles of international co-operation. Also the UN conference on the peaceful uses of nuclear energy, which met in Geneva in March-April 1987, failed to work out such principles, mainly because non-parties to the NPT refused to take account of nuclear-weapon proliferation concerns related to supplies of nuclear material and equipment.

Peaceful Nuclear Explosions. Under Article V of the NPT, the potential benefits of peaceful applications of nuclear explosions are to be made available by the nuclear weapon parties to non-nuclear-weapon parties under appropriate international observation. This promise was made in exchange for the renunciation by the latter states of the right to conduct any nuclear explosions, because there is no way to assure that a nuclear explosion has no military function.

However, there is considerable scepticism about the technical feasibility, economic viability and political acceptability of nuclear explosions for peaceful purposes. It is now recognized that conventional explosives can achieve equivalent results without the environmental and health risks accompanying nuclear detonations. As a matter of fact, there has so far been no project in which the supply of peaceful nuclear explosion services to a non-nuclear-weapon state was necessary. The USA terminated its peaceful nuclear explosion programme in 1977, and the only country which still experiments with such explosions – although it says it is willing to cease them under an overall nuclear test ban agreement – is the USSR. The prevailing opinion, at least among the parties to the NPT, seems to be that peaceful uses of nuclear explosions entail more hazards than benefits. By tacit agreement, therefore, the practical implementation of this provision has been kept in abeyance.

Disarmament Obligations. The obligations under Article VI are generally considered to be of particular consequence. For in signing the NPT the parties agreed that the self-imposed arms denial of one side – the non-nuclear-weapon states – was to be matched, ultimately, by corresponding acts of the other side – the nuclear-weapon powers. They have, therefore, undertaken to pursue negotiations "in good faith" to halt the nuclear arms race "at an early date" and to bring about nuclear disarmament. In fact, the NPT is the only existing international document under which the major nuclear powers are legally committed to nuclear disarmament. However, the nuclear arms control agreements concluded in the 1970s – the 1972 SALT Interim Agreement and the 1979 SALT Treaty – were of low disarmament value, as they merely regulated the US-Soviet competition at a high level of armaments. Besides, these agreements are no longer in force, and the validity of the 1972 ABM Treaty restricting ballistic missile defences is being jeopardized by developments in military space technology. The first meaningful measure of nuclear disarmament was taken in 1987 with the signing of the US-Soviet Treaty eliminating all ground-launched missiles with a range of 500 to 5500 km. A 50 per cent reduction of strategic missiles having a range in excess of 5500 km is under negotiation.

Whereas the sheer numbers of nuclear weapons seem to be on the decline, no effective measure has been taken so far to restrain the qualitative improvement of these weapons. In this respect a comprehensive ban on nuclear weapon testing would be essential. This is why the preamble to the NPT reiterated the determination of the parties to the 1963 Partial Test Ban Treaty (PTBT) to achieve the discontinuation of all test explosions of nuclear weapons "for all time". Indeed, without tests the development of new designs of nuclear weapons by the nuclear-weapon powers would be made practically impossible, while the modification of existing weapon designs would be seriously constrained. But the trilateral UK-US-Soviet talks on a comprehensive test ban treaty have remained suspended since 1980, and the multilateral talks on the same subject held at the Conference on Disarmament (CD) have been stalemated in a dispute over the terms of reference of the Conference working committee.

Nuclear-Weapon-Free Zones and Countries. Article VII of the NPT affirms the right of states to conclude regional treaties in order to assure the total absence of nuclear weapons in their respective territories. Two such treaties covering large populated areas have so far been concluded: the 1967 Treaty of Tlatelolco prohibiting nuclear weapons in Latin America and the 1985 Treaty of Rarotonga setting up a nuclear-free zone in the South Pacific. As an underpinning of the Treaty of Rarotonga, the Parliament of New Zealand adopted in 1987 an act establishing the New Zealand Nuclear Free Zone which comprises all land and waters within the territorial limits of New Zealand, as well as the airspace above these areas. The Act states that the Prime Minister may grant approval for the entry of foreign warships into the internal waters of New Zealand only if he is satisfied that the warships will not be carrying any nuclear explosive device upon their entry into these waters. Similarly, approval for the landing in New Zealand by foreign military aircraft may be granted only by the Prime Minister if he is satisfied that the aircraft will not be carrying any nuclear explosive device when it lands. Entry into the internal waters of New Zealand by any ship whose propulsion is wholly or partly dependent on nuclear power is prohibited [2].

In 1988, at the initiative of the opposition Social Democratic Party, the Parliament of Denmark passed a resolution requesting the Government to notify all visiting warships that they must not carry nuclear arms into Danish ports. From the formal point of view, the resolution merely reiterated the official Danish policy, which had been proclaimed more than three decades earlier, namely, that the introduction of nuclear weapons to the country is prohibited during peacetime. In practice, however, the resolution signified a rejection of the policy of "neither confirming nor denying" the presence of nuclear weapons, which has been so far strictly adhered to by the navies of all the nuclear-weapon powers. Eventually, however, under pressure exercised within NATO, mainly by the United States and the United Kingdom, Denmark agreed, as a compromise, to adopt the Norwegian formula. Norway, which has also unilaterally declared its territory to be free of nuclear weapons in peacetime, proceeds on the assumption that this declaration is respected by the visiting foreign ships or aircraft and does not seek specific assurances [3]. Several other countries as well, including members of the military alliances, have formally prohibited (as have Japan, Iceland and Spain) or have contemplated prohibiting (as has the Philippines) [4] foreign ships or aircraft to enter their territories with nuclear weapons aboard. None, however, has so far tried to enforce this prohibition.

In Sweden the ruling Social Democratic Party decided, at its 1987 congress, that efforts should be made to make the nuclear powers forgo the practice of not giving information regarding the presence of nuclear weapons on their warships. It was resolved that, should the nuclear powers decline to give up this practice, the rules for military visits would be tightened: the powers in question would be requested to make an explicit statement that nuclear weapons were not entering

Swedish territory, including its air space. The visits would be refused if no such information were provided [5]. This policy was confirmed by the Swedish Prime Minister in his speech made at the 1988 Third UN Special Session on Disarmament.

The Threatening Developments

Naval Propulsion. In 1988 it became known that India had leased a nuclear-powered submarine from the Soviet Union [6] and that the submarine was equipped with cruise missiles [7]. The lease was apparently intended to lead to the purchase of several submarines by India and/or help it to develop an indigenously designed nuclear-propulsion system. The conditions of the leasing arrangement have not been made public. However, since Soviet submarines use highly enriched uranium which can also be used to produce nuclear weapons, and since India refuses to forsake the nuclear weapon option, the compatibility of the Soviet-Indian deal with the goal of non-proliferation is open to question.

Another instance of nuclear-propulsion proliferation is the planned acquisition by Canada of a fleet of nuclear-powered submarines equipped with conventionally armed torpedoes [8]. However, as distinct from India, Canada is party to the NPT and has accepted full-scope IAEA safeguards. Consequently, the concerns regarding this acquisition are not of the same order as in the case of India. As a matter of fact, under the NPT, non-nuclear-weapon states are prohibited only from using nuclear materials for explosive purposes; the use of such materials for naval propulsion is not prohibited. Paragraph 14 of the "Structure and Content of Agreements between the Agency and States", required in connection with the NPT, provides for a special arrangement for withdrawing nuclear material from IAEA safeguards, so that it can be used in non-proscribed military activities [9]. The arrangement between the state in question and the IAEA should identify the circumstances during which safeguards would not be applied. The state would have to make it clear that the unsafeguarded material (the quantity and composition of which must be known to the IAEA) would not be used for the production of nuclear weapons or other nuclear explosive devices. According to the authoritative interpretation, based on the negotiating history, the exemption from safeguards is to be strictly limited to the material in the propulsion and should not include other stages of the nuclear fuel cycle; safeguards must again apply as soon as the nuclear material is re-introduced into a peaceful nuclear activity for reprocessing or for other, inherently non-military industrial treatment [10].

If Canada comes into possession of nuclear-powered submarines, it may avail itself of the exemption provision referred to above. This may not affect its commitment to the cause of non-proliferation of nuclear weapons, but would set an unfortunate precedent for the non-application of nuclear safeguards by the parties to the NPT [11].

Among other states known to plan the acquisition of nuclear-powered submarines are Argentina and Brazil, both hold-outs from the NPT [12].

Dual-Capable Missiles. One recommendation frequently made for the strengthening of the non-proliferation regime has been to complement the existing restraints on supplies of nuclear material and equipment by restraints on supplies of dual-capable weapon systems, that is, systems capable of delivering both a conventional and a nuclear weapon. This recommendation was partly put into practice in April 1987 when seven governments, those of Britain, Canada, France, the FRG, Italy, Japan and USA, adopted identical guidelines to control the export of equipment and technology which "could make a contribution" to missile systems capable of delivering a nuclear weapon.

The guidelines are accompanied by an annex of items to be controlled, and the items fall into two categories. Item 1 of category I covers complete rocket systems (including ballistic missile systems, space-launched vehicles and sounding rockets) and unmanned air vehicle systems (including cruise missile systems, target drones and reconnaissance drones) capable of delivering at least a 500-kg payload within a range of at least 300 km, as well as the specially designed production facilities for these systems. Item 2 of the same category covers complete sub-systems usable in item 1 systems, as well as the specially designed production facilities and production equipment therefor. All other items (3-18) in the annex are category II items. Particular restraint is to be exercised in the consideration of category I item transfers because of their great sensitivity, and there is a presumption of denial with respect to such transfers. Until further notice, transfer of category I production facilities will not be authorized. Category II items are not on a "denial" list. Almost all items in this category have uses other than for projects of nuclear proliferation concern, and their transfers will be considered on a case-by-case basis. Nuclear proliferation occupies a prominent place among the factors which must be taken into account in the evaluation of export applications.

The missile export control guidelines constitute an important initiative, in so far as they can make it more difficult, and perhaps more expensive, for countries to acquire a nuclear weapon delivery capability. However, the control regime is focused on large missiles and rockets; it is not designed to constrain more sophisticated forces. It ignores such important and relatively easily available nuclear delivery vehicles as aircraft. Moreover, the restrictions have come somewhat late. Companies from FRG, France and Italy have been collaborating for some time with Third World missile producers [13], and both the USA and USSR have provided different types of missile to several countries. In particular, Iraq, Iran, Libya, Syria and South Yemen are now in possession of Soviet-made missiles, some of which have been modified to reach a range of several hundred kilometers [14]. Negotiations are said to have been held for the sale of a new 600 km-range Chinese missile to Syria [15].

India has greatly developed its nuclear weapon delivery capability in addition to the aircraft it possesses. In particular, it has successfully tested the Prithvi short-range surface-to-surface rocket. Variously-described as having ranges of 150 to 240 kilometers, computer-guided and able to carry a payload of 1 tonne, the Prithvi has been favourably compared to the US Lance missile currently deployed with NATO forces in Europe. It is intended to become the basis for a family of missiles and is scheduled for mass-production in the 1990s; another missile under development in India is expected to have a range of 2400 km. The Indian space programme with its obvious military potential is also going ahead. India has already put a satellite into orbit, with an indigenously-produced rocket. However, in this area serious and costly engineering setbacks have occurred suggesting that the development of Indian ballistic missiles of intercontinental range is still far away [16]. Israel has demonstrated that it possesses a rocket powerful enough to launch a satellite into space [17] and has reportedly deployed intermediate-range (over 1000 km) ballistic missiles in the Negev Desert [18].

Pakistan has successfully test-fired surface-to-surface rockets capable of carrying a payload of more than 500 kg. The rockets are said to have been developed indigenously, though press reports allege that this has been done with the help of China [19].

Argentina is developing – in cooperation with Iraq and Egypt – a medium-range (800-950 km) two-stage rocket with a payload of some 350 kg [20], whereas Brazil is known to manufacture and export a wide variety of rockets, some of which are capable of delivering nuclear weapon payloads [21]. A major deal was the purchase by Saudi Arabia of Chinese surface-to-surface ballistic missiles having a range of about 3000 km, and designed to deliver nuclear warheads [22].

Some of the recipient countries may be many years away from a nuclear-weapon manufacturing capability, but if they decided to go nuclear they would certainly adapt for nuclear delivery those missiles they already possessed. It is, therefore, widely recognized that to be more effective in reducing the risk of nuclear-weapon proliferation, especially among non-parties to the NPT, the seven-nation missile technology control regime would have to be subscribed to by all suppliers of missiles, including of course the Soviet Union and China; the regime itself would have to be made sterner and its scope would need broadening. Exploratory talks on this subject have already been held between the United States and the Soviet Union [23].

The Emergent Nuclear Suppliers. The world nuclear market is dominated by a handful of industrial nations. However, the pattern of supplies is gradually changing as former buyers are acquiring the capability of developing their own nuclear technology and become themselves sellers of nuclear hardware and services. The new suppliers are mostly from the Third World. Their share in the global nuclear trade is still very modest. They are not in a position to provide modern sophisticated equipment, but they may offer an attractive alternative for those countries which shun the restrictive policies of the traditional suppliers.

So far, there has been no significant damage done to the non-proliferation regime by the emergent suppliers, because most transactions are internationally safeguarded. But the newcomers – among whom the most active are China,

Argentina, Brazil and India – may decide to be less demanding as regards the application of safeguards; they are not bound by the 1977 London Guidelines for Nuclear Transfers. As a consequence, surveillance of nuclear developments, especially in non-NPT countries, could become even more difficult. Particularly destabilizing would be an uncontrolled trade in sensitive items, including nuclear spent fuel reprocessing and uranium enrichment technologies, for it could considerably weaken the non-proliferation regime. A dialogue would, therefore, be desirable between the emerging and established suppliers with a view to working out generally acceptable rules [24].

Conclusions

In spite of certain reverses in the field of nuclear export control, the nuclear non-proliferation regime has proved to be fairly robust. There is a good chance that the next 1990 NPT Review Conference will reaffirm the validity of, and the support for, the NPT, and that the 1995 Conference, which is to decide the Treaty's future, will extend the duration of the NPT for another lengthy period.

The danger that a nuclear "threshold" country may openly join the "club" of established nuclear-weapon states continues to exist, but this danger could be greatly reduced if the regional disputes and conflicts feeding nuclear ambitions were settled. For non-proliferation has become a norm of international behaviour which cannot be easily defied. The ultimate solution to the problem of non-proliferation will, of course, be found only in a world in which the possession of nuclear weapons is universally recognized as both unnecessary and unacceptable. This goal is still remote. To bring it nearer, the process of nuclear arms reduction and elimination, initiated with the 1987 INF Treaty, must continue.

References

1 IAEA Press Release PR 88/3; *Süddeutsche Zeitung*, 26 February 1988
2 New Zealand Nuclear Free Zone, Disarmament and Arms Control Act 1987
3 *Berlingske Tidende* (Copenhagen), 15 April 1988; *Svenska Dagbladet* (Stockholm), 20, 22, 23 and 24 April 1988; *Le Monde*, 9 June 1988
4 *New York Times*, 7 June 1988
5 Swedish Social Democratic Party Congress Protocol MO- 103-gh-0010, September 1987
6 *Pravda*, 6 January 1988
7 *Le Monde*, 11 February 1988
8 Department of National Defence, *Challenge and Commitment: A Defence Policy for Canada*, (Ottawa: Canadian Ministry of Supplies and Services, 1987)
9 IAEA document INFCIRC/153
10 D. Fischer and P. Szasz, *Safeguarding the Atom: A Critical Appraisal*, J. Goldblat (ed.), (London: SIPRI, Taylor & Francis, 1985); H. Blix, "Aspects juridiques des garanties de l'Agence internationale de l'énergie atomique", *Annuaire francais du droit international*, Vol. 29, 1983

11 For a more extensive treatment of this subject, see: B. Sanders and J. Simpson, *Nuclear Submarines and Non-Proliferation: Cause for Concern*, Programme for Promoting Nuclear Non-Proliferation, Occasional Paper Two, 1988; M. F. Desjardins and T. Rauf, *Opening Pandora's Box? Nuclear-Powered Submarines and the Spread of Nuclear Weapons*, The Canadian Centre for Arms Control and Disarmament, Aurora Papers 8, 1988

12 *Jane's Defence Weekly*, 23 July 1988

13 *Financial Times*, 8 June 1988

14 *Le Monde*, 8 April 1988; *International Herald Tribune*, 2 May 1988

15 *International Herald Tribune*, 23 January 1988

16 *Neue Züricher Zeitung*, 30 March 1988; *Aviation Week and Space Technology*, 26 October 1988; *International Herald Tribune*, 4 April 1989

17 *International Herald Tribune*, 20 September 1988

18 L. S. Spector, *Going Nuclear*, (Cambridge, MA: Ballinger, 1987); *International Defence Review*, 21 July 1987

19 *International Herald Tribune*, 25 May 1988 and 6 February 1989

20 *Frankfurter Allgemeine Zeitung*, 22 December 1987: *International Herald Tribune*, 20 September 1988; *The Guardian*, 21 September 1988

21 A. Karp, "The frantic Third World quest for ballistic missiles", *Bulletin Atomic Scientists*, June 1988

22 *International Herald Tribune*, 19 - 20 March 1988; *Le Monde*, 7 April 1988

23 *New York Times*, 27 September 1988

24 For detailed proposals to this effect, see L. A. Dunn, "Nonproliferation: The next steps", *Arms Control Today*, Washington, November 1987

The Adequacy of IAEA Safeguards for the 1990s

*Michael J. Wilmshurst**

Apart from IAEA safeguards, there are also others: Euratom safeguards, and national safeguards systems. This paper discusses only IAEA safeguards.

Before considering their adequacy for the 1990s, how do IAEA safeguards look in 1989? The 1988 Annual Report will show that as of 31 December 1988:

1. There were 168 safeguards agreements in force with 99 states;
2. Safeguards were being applied in 57 of these states; in 41 under agreements arising from the NPT, in one under the Tlatelcolco Treaty, in eleven under specific agreements not connected with any Treaty and in four nuclear weapon states.
3. Voluntary offer safeguards agreements were in force in four of the nuclear-weapon states; an agreement with a fifth had been concluded but had not come into force;
4. 514 nuclear facilities were under safeguards or contained nuclear material; 407 locations outside facilities contained small amounts of safeguarded material, and two non-nuclear installations were safeguarded. In total there were 923 installations in 57 states which were under safeguards or which contained safeguarded material.
5. The approximate amounts of nuclear material under Agency safeguards included:
   - 10.8 tonnes of separated plutonium outside reactor cores,
   - 0.5 tonnes of recycled plutonium in fuel elements in reactor cores,
   - 254.4 tonnes of plutonium contained in irradiated fuel,
   - 13.1 tonnes of high enriched uranium,
   - 31,704 tonnes of low enriched uranium.

   In addition, safeguards are applied to 1454 tonnes of heavy water in those countries (basically non-NPT countries) where safeguards are applied to non-nuclear material as well as to nuclear material.
6. To do this work the Safeguards Department of the Agency employed 516 persons, and operated a budget of $49.5 million dollars, representing 33% of the Agency's overall budget for 1988.

Were these safeguards adequate? In 1988, as in all previous years, the Director General was able to state in his report on 1987 that:

* Although this paper has been written with the help of colleagues in the IAEA's Secretariat, it reflects the author's personal views and is not a statement of official IAEA views.

"The Secretariat, in carrying out the safeguards obligations of the Agency, did not detect any anomaly which would indicate the diversion of a significant amount of safeguarded nuclear material – or the misuse of facilities, equipment, or non-nuclear material subject to safeguards under certain agreements – for the manufacture of any nuclear weapon, or for any other military purpose, or for the manufacture of any other nuclear explosive device, or for purposes unknown. It is considered reasonable to conclude that nuclear material under Agency safeguards in 1987 remained in peaceful nuclear activities or was otherwise adequately accounted for."

It is my expectation that in June of 1989 the Director General will make a similar statement about 1988.

Although there are many different types of Safeguard Agreements, they can be divided into two basic types: those which apply safeguards to all nuclear material in a country, and those which apply safeguards to a specific item of nuclear equipment and/or to a specific quantity of nuclear material. The first type, often known as full scope or comprehensive, includes the agreements concluded in accordance with the NPT.

With that as a background I return to the title of the paper – The Adequacy of IAEA Safeguards in the 1990s.

If that title means the adequacy of IAEA safeguards to confirm that there has been *no* proliferation in the 1990s, I can give a short comment. IAEA safeguards are not now and will not be in the 1990s adequate for that purpose. For such time as there are states – whether nuclear-weapon states or non-nuclear-weapon states – which have not placed all of their nuclear activities under safeguards, it would be impossible for the Agency to confirm that there had been *no* proliferation.

The Agency can only do what it is asked to do by its Member States. It can only work within the limitations of the Safeguards Agreements concluded with states. It is for this reason that the annual statement by the Director General refers to *safeguarded* nuclear material, and to "facilities, equipment, or non-nuclear material *subject to safeguards*".

So, the Agency's safeguards will not be, and cannot be expected to be, adequate for a task which the Agency has not been asked to perform. Will they be adequate for the tasks that it *has* been asked to perform?

I would like to give an automatic assurance that they will, but instead I shall outline a number of areas of potential difficulty – or of present difficulty – which taken together, might limit the adequacy of Agency safeguards.

I shall divide these areas of potential difficulty into seven headings: Scope, Cooperation, Impartiality, Credibility, Duration, Financial, and Technical.

Scope. Agency safeguards can only apply to nuclear material or installations covered by Safeguards Agreements. As I have explained, some provide full scope safeguards; some do not. It is not for the secretariat of the Agency to try to influence states to accept one type of Safeguards Agreement rather than another. However, I think that it is a legitimate concern of the secretariat to note that the task of safeguarding certain installations becomes more difficult in states where there exist other, unsafeguarded, installations. The existence of parallel nuclear industries makes it hard to confirm that there has been no diversion of nuclear material from the safeguarded to the unsafeguarded installations. Another difficulty, without full scope safeguards, concerns whether and how to apply safeguards to technological information. For these reasons, and not for any political reasons, I feel that safeguards would be more adequate if there were more full scope agreements and fewer unsafeguarded installations.

Another aspect of scope concerns the safeguarding of nuclear-weapon states. If, as the 1985 NPT Review Conference suggested, safeguards were to be applied to all peaceful nuclear facilities in the weapon states, the size of the Agency's safeguards budget would have to be doubled or even trebled, with a corresponding change in the size and balance of the secretariat as a whole.

Cooperation. It would be too much to expect governments and operators to feel very hospitable towards international inspectors, but most of the Agency's member states seem to tolerate safeguards as something that they have accepted as serving a necessary, if tiresome, purpose. Unfortunately some member states are less philosophical, and a few of them seem to resent or even, in extreme cases, to obstruct the operation of safeguards. Examples range from delays and refusals connected with the designation of inspectors to more serious incidents, such as refusal to take samples out of the country and reluctance to agree to the installation of new surveillance or monitoring equipment, together with difficulties over safeguards costs and liability responsibilities. These examples rarely betray ill will; more often they reflect concern for national or commercial security. But whatever the motivation, poor cooperation in such areas limits the smooth running of the safeguards system. I should also stress that most governments and operators are aware of the purpose of safeguards and of the need for them. Several industry representatives have told us that without the Agency safeguards system there would be little or no industry, and little or no nuclear trade. But not all governments and operators seem to feel that way.

Impartiality. Under this heading I am concerned less with relations between member states and the secretariat, and more with the relationships between member states. Suggestions have been made in some publications that, because for many years the Agency's General Conference has passed resolutions criticising them, the IAEA is hostile to Israel and to South Africa – or, more politely, that it has been affected by "extraneous political issues". Such accusations may be little more than political point scoring when made, but they can be revived later to damaging effect. As unhelpful is the traditional conflict between those – mostly industrialized – countries who see safeguards as the prime purpose of the Agency and those – mostly developing – countries who see technical cooperation as the Agency's main role. These differences are paraded mostly in connection with the size and priorities of the Agency's programme and budget, but they affect attitudes more generally and provide lobbies for and against expenditure on safeguards. If, as a result, there were doubts about the motivation of the supporters of the Safeguards Programme, that programme could suffer.

Credibility. To be accepted positively, by operators and by governments, safeguards must be credible. If credibility is lost there is little point in continuing to spend all of those millions of dollars. Recent attacks on safeguards credibility have been based upon press misunderstanding of the purpose and the limitation of safeguards, and have been based more specifically upon leaked copies of the annual Safeguards Implementation Report. The secretariat has reviewed the way that that report is written. We hope that future editions will be written in a less specialized style, with a broader audience in mind, and that there will be less chance of criticisms being based upon misunderstandings.

But the best way of maintaining credibility is by maintaining efficiency. If the safeguards system is seen to work, operators and governments are more likely to give it their full support.

Duration. The safeguards system is based upon a variety of types of Safeguards Agreements, but the type of agreement most commonly used is that concluded in accordance with the NPT. These agreements all contain the provision that they will remain in force "as long as the State is Party to the Treaty on the Non-Proliferation of Nuclear Weapons".

The NPT in its turn, provides that the Review Conference in 1995 shall decide "whether the Treaty shall continue in force indefinitely, or shall be extended for an additional fixed period or periods".

If the decision were to be an extension for a fixed, but short, period of – say – one year, the result would be that in 1996 the Treaty would no longer be in force, and the majority of the Agency's Safeguards Agreements would also no longer be in force.

There are other scenarios. It is possible that the parties to the NPT will decide to replace the present Treaty with another, improved, version. There is no guarantee that in such circumstances all of the present parties would adhere to the new Treaty or renegotiate their Safeguards Agreement. It is also possible that a majority decision to extend the NPT could cause such offense to the minority who voted against extension that they would decide to exercise their right to withdraw. Again, the result would be the expiry of the Safeguards Agreements.

None of these may happen. Quite possibly the 1995 Conference will decide to extend the Treaty indefinitely: But it is worth bearing in mind that, unless some special arrangements are made if the Treaty is not renewed, most of the Agency's Safeguards Agreements stand or fall with the Treaty. If they fall the effect upon nuclear trade and upon the non-proliferation regime would be chaotic.

Financial. Here we encounter the policy of Zero Real Growth. It was invented by some of the states whose individual assessment is more than 1 per cent of the budget of the UN and the International Organizations.

It was, originally, meant to improve efficiency and to impress upon the organizations the need for careful management and close budgetary control. The aims were good, and the initial effects were good. Unfortunately, the policy of Zero Real Growth – following upon an earlier policy of Near Zero Growth –

became an end in itself. No matter what additional demands were placed upon the secretariat, the budget could not expand to meet those demands.

Within the IAEA this exacerbated the existing differences between those member states whose main interest was in safeguards and those whose main interest was in the Agency's promotional activities. As the cake could not be made bigger, views differed as to how to cut the slices. In earlier years, the supporters of safeguards had secured agreement to increases in the Safeguards budget by agreeing to increases in promotional activity. This seems no longer to be possible, and the Safeguards budget has had to accept very small or no increases in real terms.

At the same time more installations have come under safeguards, the amount of nuclear material has increased, and technical improvements in installations and in safeguards techniques require greater expenditure.

Some of the increased expenditure has been met by greater efficiency, some by assistance from National Safeguards Support Programmes. However, if the Zero Growth policy continues it seems quite possible that safeguards will no longer be adequate throughout the 1990s, simply because there will not be the money to pay for what is needed to maintain the credibility of the system.

Technical. Two developments will make the application of safeguards more difficult technically. The first is qualitative; enrichment technology is being developed in an increasing number of states. This means that it is becoming more difficult to differentiate between direct use uranium – enriched above 20 per cent U235 – and non-direct use uranium – enriched below that level. The basis for many safeguards approaches may lose their validity. The second problem is quantitative; the constantly growing stockpile of separated plutonium requires very intensive and frequent inspection, as does the increasing quantity of plutonium in cooled and easily reprocessable, spent fuel. Other developments, such as laser enrichment or large remote controlled reprocessing plants, will make safeguarding a more complex and a more expensive process.

Conclusion. My conclusion, after considering these issues, is that although the IAEA safeguards system is adequate for its present task, there is no certainty that it will continue to be adequate during the next decade. Its adequacy will depend upon greater cooperation from governments, not only in the mechanics of implementation but also in the provision of sufficient resources; its adequacy will also depend upon the size and nature of any additional task given to the Agency. If, for example, the nuclear-weapon states were to agree to place all their peaceful nuclear installations under safeguards on the same basis as the non-weapon states, or if nuclear material now used in warheads were to be transferred to the civil nuclear cycle under the terms of a disarmament agreement, the Agency's task would expand, and its resources would need to expand with the task.

This is a necessarily cautious conclusion. It should be seen against the background of a unique international system of intensive inspections and detailed accounts. The world community which created the safeguards system is unlikely to allow it to lose its credibility for want of resources or lack of cooperation at a time when verification has become a topic of current concern, and when the Agency's safeguards system is seen as a possible model for other, newer, systems. The new verification systems may not follow the exact example of the IAEA, but their acceptability will be largely based upon the international acceptance for thirty years of IAEA safeguards. It is, perhaps, the wider importance of verification in the 1990s which makes it seem unlikely that IAEA safeguards will prove to be anything but adequate.

The Nuclear Trade Regime: A Case for Strengthening the Rules

Harald Müller

Why Nuclear Trade Restrictions Still Matter

Some see nuclear trade restrictions as a waning asset. After all, technology is spreading, and thus curbs on nuclear exchanges have lost their meaning. This is not quite true. First, there are still many states for which it is not conceivable to develop nuclear weapons without outside technical assistance. Second, even those with indigenous capabilities could accelerate the speed and enhance the quality of their military programmes with foreign inputs. In either case, delay may buy needed time to manage or to solve the conflicts that motivate governments to seek nuclear capabilities. Third, as an indispensable element of a mutual confidence-building non-proliferation regime, exports must be covered by clear and comprehensible rules by which both exporters and importers abide. However, export restrictions have met objections from both sides.

Objections to Curbs on Nuclear Trade

Developing countries have constantly objected to agreements among the exporters. In their declaration of 1985, the foreign ministers of the nonaligned countries have criticised the industrialized states for "raising obstacles" and "fixing conditions which are incompatible with the sovereignty of the developing countries" [1]. A dispassionate look at these accusations, however, puts their realism into doubt.

First, the main problems of trade restrictions among NPT parties emerged between the suppliers themselves rather than between them and Third World importers; the consequences of the American Nuclear Non-Proliferation Act for Japan, the EC and Switzerland, and conditions attached to the uranium exports by Canada and Australia were, in terms of economic implications, the most consequential constraints laid on NPT members, and they hit the industrialized users of nuclear energy, not developing countries.

Second, there are five well-known cases of transfer restrictions imposed on developing countries party to the NPT [2]:

- The USA dissuaded France from supplying reprocessing technology to South Korea in the mid-seventies. France cancelled the sale only after having been shown evidence that South Korea was then seriously considering a military nuclear option. Technology denial was thus justified.

- The same applies to the forceful US dissuasion campaign against reprocessing in Taiwan, first in the seventies, and again very recently. In both cases, the Taiwanese projects had such unambiguously military overtones that US pressure was completely justified.
- The USA denied Iran fuel for its research reactor, and West Germany would not allow Kraftwerk Union to complete the Bushehr reactor as long as the Gulf war continued. In the first case, the USA had good non-nuclear reasons for an unfriendly policy; the fate of the hostages, taken against all rules of international law, is not forgotten. For the Germans, is was unacceptable to contribute to an even more dangerous situation in the Gulf, after Iraq had clearly demonstrated its determination to attack the construction site at Bushehr. Besides, the conduct of the Khomeini regime with respect to its obligations under international law makes it imprudent to supply nuclear technology in any case.
- Iraq was unable to obtain from France or any other supplier an exact replication of Osiraq. Given Iraq's blatant breach of the Geneva Protocol, nuclear suppliers would be crazy to send a facility well-fit for plutonium breeding, fueled by HEU, to Mr. Sadam Hussein.
- Libya – apart from the Tajoura nuclear research centre reportedly under firm Soviet control – did not get power reactors or other basic nuclear support from either the Soviets or Western countries. This policy of restraint is only wise in the light of Mr. Khadafi's connections with international terrorism and his various breaches of international law, including presumably the use of poison gas in Chad.

Third, apart from these cases, only non-members of the NPT were victims of denial, most prominently Pakistan, South Africa and Israel. We have still to see a case of a faithful and credible NPT adherent being denied access to nuclear technology. At the moment and in the foreseeable future, the obstacles will rather be either lack of electricity demand and the availability of cheaper primary energy resources, as for Mexico, or financial difficulties, as for Egypt.

Thus, the continuing protest against suppliers' policy uttered by the G-77 is, by and large, a submission of faithful Third World NPT members to their own abuse by either untrustworthy Treaty adherents or non-members. India and Argentina have been most adept in steering the mass of developing countries towards a course that is apparently critical of industrialized suppliers, but in fact directed against the NPT as such [3]. It is a great pity that developing countries party to the NPT have found neither the courage nor the energy to set up a caucus of their own, to liberate themselves of the self-serving machinations of non-members, and to define, within the terms of their Treaty membership, their own criticism as well as their demands towards NPT suppliers. This, in fact, would put far more pressure on NPT suppliers to consider either more technological assistance or greater cooperation in defining the terms of supply than general G-77 complaints.

It is worth recalling that there have been second thoughts on nuclear export curbs even among industrialized countries, too. After all, Article IV and the explicit restrictions it puts on the applicability of Article III; the restrictive rules for safeguards (material only, no facilities, access rights for inspectors only to strategic points, etc.) are all results of the insistence of advanced countries. The application of full-scope safeguards export policy was blocked by a group of industrialized countries, and still is [4]. The reasons are manyfold. The desire to enhance exports; mistaken fears that safeguards would impede the growth of domestic nuclear industry; misplaced concerns of national sovereignty; misgivings over American unilateralism; and sincere concerns that a strict export policy would poison the international climate and diminish prospects for integrating the present hold-outs in the regime – all these considerations worked together to create and maintain the aforementioned attitude and have been a stumbling block for the effective implementation of export policies in several countries. Proposals for strengthening export controls and establishing further constraints on exports will thus meet a rather powerful coalition of opposition.

The Legitimacy of Trade Restrictions

By their membership in the NPT, states undertake not to contribute to another's nuclear weapons development (Art. 1) and to have all nuclear exports properly covered by IAEA safeguards (Art. 3.2). For this purpose, member states must establish systems of nuclear export regulation and control. According to Article 3, those systems should cover not only fissionable material, but also facilities, technology and components which may be used to produce or process such material. While exporters' obligations with respect to fissionable material are elaborated somewhat in INFCIRC/153, the same is not true for the other items. It was largely left to the member states to consider how to deal with the subject. Two international groupings with overlapping membership have tackled the issue: The Zangger Committee as a consultative body of NPT supplier states, and the London Group as a NPT-independent entity which helps France to support the NPT without openly admitting it.

A main weakness of this situation is that nuclear trade rules are agreed to only among a minority of NPT member states, overwhelmingly industrialized states. This is a twofold shortcoming: The apparent ganging-up of the "haves" lends some credibility to the allegation that non-proliferation is being abused for denying technology to developing countries; and the fact that many states stay outside the nuclear trade regime may create dangerous loopholes.

It may appear that this is not all that meaningful; the great majority of the absentees is in no position to supply anything nuclear to anybody. Yet this argument is short-sighted. First, among those absentees, there are some with an incipient technological capability which, under the pressure of economic need and in the absence of an export control system, may be abused: somebody from outside could well prescribe the specifications of desired items to be manufactured in this country. Second, each country with a sizable trading company could be utilized as intermediator. Third, each such country can be employed as a transit area, thus circumventing export constraints embedded in "country lists" in the advanced exporting states. It would, therefore, mean great progress to broaden the membership of the nuclear trade regime.

One has to distinguish between four layers of legitimate export restrictions:

1. Conditions attached to nuclear exports to all recipients; they include peaceful-use statements, guarantees that IAEA safeguards will be applied to the items supplied, reasonable physical protection and the application of the same conditions in the case of reexport.
2. The denial of sensitive exports to all countries if the desired item makes no sense in the context or at the stage of a country's civil nuclear programme. This policy is, of course, heavily contested. It presents the clash of two valid political principles: the sovereign right of countries to decide on their domestic energy programmes; and the sovereign right of countries to determine how to follow their non-proliferation commitments. There are three ways to assess the sincerity of a country's non-proliferation commitment: the legal undertakings of a country; its past record concerning its legal commitments; and the economic sensibility of a nuclear request. To be on the safe side, all three criteria must be met; it makes no sense to supply reprocessing technology to somebody who does not even operate a single LWR; it is likewise imprudent to deliver huge amounts of HEU to a country which operates a medium-sized research reactor and asks for a large fuel stockpile. In both cases, the request may be innocent and due to inept planning – but who knows? Nuclear suppliers are well advised to say "no" in these cases. Article IV of the NPT forces them to conduct nuclear cooperation only "to the extent possible". If a nuclear export could threaten the other main duty of an exporter, it is appropriate to consider it "not possible" under the terms of Article IV.
3. The same principle also applies to supplies to "non-faithful" members, even in a more extensive sense. The international community must face the facts: a few countries party to the NPT cannot be trusted. Past actions, either direct attempts to acquire nuclear capabilities, or even public announcements of their intention to do so, or clear evidence of breach of other international legal obligations – such as the Geneva Protocol – legitimize a reconsideration of Article IV obligations. No cheater must be permitted to abuse its Treaty membership as a convenient cover for secret nuclear-weapon programmes. Enhanced attention by supplier states for the acquisition efforts of this handful of states is thus fully in agreement with their NPT obligations. It should be thoroughly considered if any nuclear cooperation with those suspects, even in non-sensitive areas, is advisable. In principle, all knowledge, technology and know-how gained in the civilian sector can be diverted for military purposes.

4. Lastly, there is the group of non-members of the NPT. The full-scope safeguards debate is well-known, and the main arguments pro and con are on the record. The basic dilemma here is how to deal with new suppliers [5]. If one denies all cooperation to them, would they feel compelled to offer nuclear supplies safeguards-free out of "defiance"? Of course, it is not immediately comprehensible what interest they would have in further nuclear proliferation even as non-members of the NPT, and their export behaviour so far is very reasonable. Yet, given the record of early French export policies from the mid-fifties to the early seventies, the defiance motive cannot be completely excluded. On the other hand, it is a worrysome thought that ostensibly peaceful cooperation, even within the boundaries of export controls and safeguards, could foster military ambitions in the recipient country. For example, in 1978 West Germany supplied minute quantities of uranium hexafluoride (UF_6) to Pakistan; licences were obtained, and the agreement of Euratom secured. Yet, at the same time, Pakistan acquired illegally an uranium conversion plant from a West German engineering firm and was secretly purchasing the components for its military enrichment plant. Not much fantasy is needed to imagine that the small UF_6 import served to train Pakistanis in handling the material and, therefore, accelerated their proficiency for conducting the military programme.

Another example: In the West German-Brazilian deal, a replication clause was properly introduced, which obliged the Brazilians to accept safeguards on all facilities based on transferred technology, even if indigenously built, up to twenty years after the transfer. The jet-nozzle pilot enrichment plant is completely safeguarded. But Brazilian technicians learned a lot of enrichment know-how in working for and in that plant, e. g., high vacuum techniques or UF_6 behaviour under high pressure. Reportedly, there has been a high turnover of personnel from the jet-nozzle to the "parallel", military-controlled ultracentrifuge programme.

For these reasons, nuclear cooperation with non-NPT members should be kept to a minimum except in the most extraordinary circumstances; large contracts should be excluded, and small supplies granted only after the most careful scrutiny. One area where cooperation is recommended despite these considerations is nuclear safety. Humanitarian considerations and the risks of trans-border pollution in case of an accident outweigh the security aspects in this area.

Strengthening Export Control Policy

Many states have created some system of export controls. Yet experience and revelations in the past five years show that these controls are far from being perfect; illegal exports and the utilization of existing loopholes have contributed to unsafeguarded nuclear activities in several countries [6]. In this section, steps to close the gaps in the international nuclear trade control network will be discussed.

The success of export control policy depends on the interplay of three variables: the coverage of law and regulation, the probability of detection, and the sanctions against violations of the law.

Coverage of Law and Regulation. Comprehensive coverage of law is essential in order to prevent the emergence of loopholes and their prompt utilization by clever and inventive entrepreneurs. History teaches that the following loopholes must be dealt with:

- The *decomposition* of facilities and components. Law and regulation must cover not only the "whole" of an export item, but also the conscious decomposition – conducted nationally as well as internationally – into parts and subcomponents.
- The export of *items slightly below agreed specifications* but which could be used, with an acceptable loss of efficiency, for military purposes. Zirconium, graphite, beryllium, maraging steel and so on are usable in this way even if the supply does not exactly meet the specifications spelled out in the Zangger list or the London Guidelines. It would be wise to deny those items to certain countries with a considerable quality margin below the agreed specifications.
- *Technology transfers*: Not all regulations concerning this area are satisfactory [7]. Yet this is one of the most pressing issues in international nuclear trade. The incipient industrial capability of several countries may suffice, with some efficiency loss, to build the items and facilities needed for a military nuclear programme. The lack of know-how, however, particularly in the engineering sector, is still a strong inhibiting factor. To subject technology transfer – models, blue-prints, computer programmes, disks and so on – to the same rules as hardware exports is essential.
- *New Technologies*: Allowance should be made for the controlling agencies to deny licences for doubtful exports, pending a quick investigation, even if the items concerned do not appear on any official lists. For some newer technologies – e. g. laser isotope separation – the items to be controlled are not well-spelled out. New technologies may appear, and the procedure of amending existing lists are cumbersome. Thus, dangerous exports should be prevented as they emerge. This relates also to the issue of nuclear-powered submarines, its reactors, components, and fuel. The issue has not yet found a satisfactory solution in terms of export agreements and safeguards regulations. It is distinct from other export control problems as the items involved are unambiguously military. Yet in order to keep the regime as a whole intact, coverage of this issue is highly desirable [8].
- *Transit*: The use of a destination given on an export licence with the intention to re-export immediately from this destination to another customer is one of the most dangerous loopholes existing. National law must thus subject all transit of listed items to licence requirements, and must establish the necessary controls at ports and airports.
- *Transit trade transactions*: This is the same problem in even more pressing form: re-export from a place before customs clearing, or change of destination

in flight or at sea. The country where the trading company resides must establish legal rules to subject such transactions to licensing requirements [9].

- *Activities abroad*: It should be made unlawful for any citizen to participate in unsafeguarded activities in a foreign country. Nuclear (or chemical) mercenaries should be outlawed by common agreement of all faithful NPT parties. No one having worked for a Khadafi or Khomeini bomb should be permitted to enjoy the profit of his work in Florida or at the Lake Leman. By implication, holdings must be held responsible for the related activities of their subsidiaries abroad. It should be clear at the outset that the coverage of this stipulation should be narrowed to weapons of mass destruction. As a West German, I am well aware of the possibility of mindless extraterritorial application of national law; West Germany, among others, was a victim of an ill-conceived US campaign of economic punishment during the dispute on the East-West natural gas pipeline in 1982. Each agreement of this kind is thus contingent on iron-clad assurances, lest it not be abused for second-class foreign policy purposes or even for crude protectionist aims.

Detection Probability. The likelihood of discovering an intention to cheat, or at least to catch the perpetrators in or after the act is an essential part of deterrence. This probability, in turn, depends on five factors:
Adequacy of licensing process: This means the establishment of due procedures for licensing, the flexibility of the licensing authority to apply different criteria and variable depth of scrutiny to exports depending on the sensitivity of the items and their destination, and the appropriate involvement of foreign offices and intelligence services. An essential precondition is that economic and energy ministries and atomic energy commissions are not at liberty to decide on critical licences by themselves without the consent of foreign offices, prime minster's offices, or defence ministries.

The licensing process needs highly qualified staff in the agencies involved in sufficient quantity to do the job. Licensing agencies must be afforded the necessary technical capabilities, in terms of modern data processing equipment and direct communication links with all offices involved. On-line access to relevant intelligence data and the establishment of liaison staff between licensing agencies, customs control, investigative offices, intelligence, and foreign ministries, would no doubt enhance the depth of scrutiny and, at the same time, speed up the licensing process in the interest of legitimate industry.
Quality of customs control: It is, of course, impossible to employ masses of nuclear engineers as custom agents. However, there are a couple of steps that can improve detection capabilities at the margin. Younger customs agents should be offered technical training courses to alert them to indications of suspicious transactions. By concentrating this technically trained staff, first at the most crowded export places (main airports and ports), later spreading them to less prominent spots, would, at relatively low cost, enhance the capacity of customs staff to blow the whistle just in case. Second, a small specialized nuclear customs staff should be held at the customs control centre; this staff should be available for on-the-spot

random checks (thereby providing a residual, low-probability deterrent for potential perpetrators) as well as for assistance to local whistle-blowers. Third, the special nuclear investigative staff of customs agencies should be strengthened to support other state authorities in investigations. Fourth, the communication links of customs agencies to foreign services should be strengthened. Fifth, customs agencies should share with licensing offices and intelligence services a high-quality data system on the nuclear export capabilities of national and foreign companies.

System of information on industrial capabilities: As mentioned before, one of the essential ingredients of an export control system is a clear overview of the responsible authorities over the skills of companies in the nuclear field. A register of all firms licensed to handle fissionable material; a register of all companies with a competence in designing, engineering and constructing nuclear facilities or key components; lastly, a register of "potential" participants (measured by their skills in special tools and material handling, for example) would enable export offices to discover "suspicious" licence applications; customs offices would be enabled to investigate as soon as well-known broker companies (the notorious Pakistani and Iraqi firms, for example) approached domestic agents, or when a company listed in the register was particularly active in a suspicious country.

Instruments of investigation on companies' activities: This starts, as mentioned, with the availability of competent special staff and a comprehensive data system in the investigative offices. Beyond that, investigators must have far-reaching intrusive authority – in accordance with the general rules and due procedures of law – in order to inquire quickly and effectively and to prevent all attempts at cover-up. They need immediate and unimpeded access to all books, records, data, rooms and materials of a company, the right to question company staff, under oath, if necessary; to seize all documents, seal all rooms as required, and to detain suspects.

Sharing of information and intelligence among states: Perhaps the most essential condition for effective export controls is international cooperation. Intelligence sharing has already helped a lot in apprehending violators of export laws. In fact, a majority of the cases discovered in the FRG was due to information from abroad. The sharing of such information is essential because the quality of information is asymmetrical in different states. Some states have no efficient foreign intelligence at all or very little while others are not focusing on economic transactions. But the need for information does not stop here. First, as explained above, transit and transit trade transactions pose special challenges to authorities. Handling this challenge can be greatly facilitated if countries of origin routinely inform the appropriate transit agent offices in the recipient country of a transaction occurring. Likewise, the net becomes far more impenetrable if the recipient countries inform countries of origin that they have obtained information on a certain transaction. The status of International Import Certificates should be regularly exchanged; this would inform countries of origin if some export did not reach the stated destination, and thus could trigger timely investigations.

The international accumulation of information would help with the "small quantity problem" [10] which is a troublesome issue in international export controls. If small quantities of plutonium, or heavy water, or any other sensitive material are exported from a couple of places to the same destination, several critical quantities could easily build up; this could be prevented by the immediate notification of the IAEA of such exports. The Agency could then ring the alarm if there was an accumulation of such transactions.

Similarly, the decomposition of critical facilities and components in an international setting would be impossible if a central data net would immediately inform all potential suppliers of partial transactions conducted anywhere. Lastly, mutual information about the capabilities of mother companies would enable countries in which their subsidiaries are located to properly assess their skills in the nuclear sector and thus complement the proposed register accordingly.

Sanctions. Sanctions against violations of nuclear export laws are an important part of the whole non-proliferation system. The record of punishment of violators so far has been dismal. The most important case, the conviction of Albert Migule in West Germany, ended in a DM 30 000 fine (for a contract earning DM 15 million in sales) and an eight-month suspended prison sentence [11]. It is quite clear that sanctions like this are not appropriate means of deterrence.

The *first* important principle is that all aspects of a possible illegal activity must be covered. Neglect must be punishable as well as intent; the attempt should be equally subject to punishment. Likewise, the prosecutors should not be burdened with the proof that state interests, peace, or some other standard were actually violated; the evidence for a threatened violation should suffice for imposing a sentence.

Secondly, it is essential to enlarge the circle of responsibility as much as possible in order to engage the interest of a large group of persons in preventing illegal actions. Mother companies should be held responsible for the activities of their subsidiaries. Companies as a whole should be liable, not just individual employees. In the case of employees, superiors with supervisory duties should be sentenced for neglect if their subordinates are found guilty.

Thirdly, the sentence itself should be commensurate with the possible damage expected from the incriminating export. Unfortunately, the prescriptions for punishment in most supplier states are in the tradition of trade law; punishment in trade law is measured by the standard of mercantilistic-protectionistic interest, with the intent of strengthening one's own national companies. As a consequence, fines and sentences for the violations of nuclear export rules are equivalent to the illegal import of some textiles. The qualitative difference between the illegal acts is in many cases not properly taken into account. Punishment often amounts to little more than a tiny addition to the value-added tax, a penalty companies are often ready to incur to gain a huge profit.

Thus, changes in the following direction are recommended:

As a rule, the profits from illegal exports should be confiscated. In addition, a fine should be imposed that is commensurate to the size of the illegal sale as well as to the size of the company.

Licensing agencies should be obliged to reconsider the granting of all licences in the nuclear field to the company in question. As a rule, all such licences should be revoked and no new licences granted. An exemption to this should be subject to interministerial consultation, involving at least the foreign and defence ministries.

At the personal level, violations of nuclear export laws should not be treated as a cavaliers delict. If a state has decided that nuclear proliferation is not in its national interest, breaches of the relevant regulations should be punished as high treason. It can be expected that the high personal risk which the expected punishment involves will be adequate to prevent all such violations from occurring when more than a very few people are involved.

At the international level, the violation by a company of nuclear export laws in one country should entitle other countries to revoke that company's free trade privileges. The details of such a rule should be worked out very thoroughly; its abuse for protectionist purposes cannot be excluded. Furthermore, the country of residence of the respective company may feel a need to postpone, suspend or waive law enforcement if it fears great foreign trade disadvantages. On the other hand, highly negative international economic consequences would work as a mighty incentive for the company to stay within the boundaries of law, and for the country to prevent any violation.

Fourthly, an importing country that encourages citizens to violate their country's domestic law should not get away unscathed. Economic sanctions should not be ruled out against the profiteers of export law violations [12]; in most cases, the economic stakes of exporters in the country concerned are small enough.

Conclusion

Greater international harmonization of nuclear export law, regulation, and even the rules of punishment are badly needed. The harmonization should go beyond the present supplier groups. As the discussion of transit trade transaction has shown, all trade poses a certain risk. It would be wise to try to create a common code for all faithful NPT members. Why not agree on something like the CAS rules, [13] amended by detailed common rules for export control? Developing countries will justifiably ask for support in creating the respective agencies; industrialized countries should not hesitate to provide adequate funding. Moreover, since for a lot of countries the creation of national offices would be both expensive and unjustified given their inability to export anything meaningful by themselves, a small but capable international export control agency could play a helpful role. It could serve as a nuclear export data centre; its well-trained staff would be at the disposition of smaller member states whenever their governments, by their own sovereign decision, determine that assistance was needed for some investigation. Such an agreement among the NPT countries would terminate the illogical and counterproductive alliance between faithful members of the Treaty and those with a doubtful non-proliferation posture. Most likely, developing countries would ask for some offer of goodwill – in terms of enhanced technical assistance – in return. Given the high stakes, this sacrifice should be worth it to the industrialized world.

In the end, no export control system will be effective without a serious commitment from the highest echelons of leadership. Nuclear non-proliferation must be established as a state goal, as a part of national security, and superordinated to foreign economic policy aspects.

If this is not the case, experience shows that even sensible law and regulation will not prevent violations nor catch the violators. Indifference by the political leadership will leave export control agencies understaffed and underfunded, and will facilitate the emergence of symbiotic relationships between the regulators and the regulated. Good laws are of no use if implementation is lousy. In such a climate, as West Germany's experience shows, everything is possible. To have foreign offices and defence ministries involved puts a useful brake on export licensing. Yet permanent caution will only prevail if ministers, or even heads of state, pay regular and visible attention to the issue. Well-informed investigative journalism, regular scholarly focus on the issue, and public pressure groups devoting their energy to nuclear non-proliferation – all this will remain necessary to persuade leading politicians to keep their eyes on curbing the spread of nuclear weapons.

References

1 See the OECD/NEA report, *The Regulation of Nuclear Trade, Nonproliferation, Supply, Safety*, Vol. 1: *International Aspects*, (Paris: NEA, 1988) p. 132

2 See Leonard S. Spector, *Going Nuclear*, (Cambrigde, MA: Ballinger, 1987)

3 Lawrence Scheinman, *The International Atomic Energy and World Order*, (Washington: Resources for the Future, 1987) pp. 191–92, 246, 218–24, 268–70

4 See Harald Müller, "Nonproliferation Policy in Western Europe: Structural Aspects", in H. Müller (ed.), *A European Nonproliferation Policy: Prospects and Problems*, (Oxford: Univ. Press, 1987) pp. 85–92

5 On the "new suppliers" issue see various useful articles in Rodney W. Jones et al. (eds.), *The Nuclear Suppliers and Nonproliferation*, (Lexington, MA: Lexington Books, 1984); Lewis A. Dunn, *The Emerging Nuclear Suppliers: Some Guidelines for Policy*, CISA Working Paper No 61, (Los Angeles: Centre for International and Strategic Affairs, UCLA, 1988)

6 On illegal exports see Leonard S. Spector, *The New Nuclear Nations*, (New York: Vintage, 1985) pp. 22–39; Gary Milhollin, *Germany's Heavy Water Laundry*, (Washington: Wisconsin Project on Nuclear Arms Control, 13 June 1988); Harald Müller, *After the Scandals: German Nuclear Nonproliferation Policy*, (Frankfurt: Peace Research Institute Frankfurt, 1989)

7 This conclusion is derived from a comparative analysis of Vol. 2, *National Regulations*, OECD/NEA report, op. cit.

8 On the nuclear submarine issue see Ben Sanders and John Simpson, *Nuclear Submarines and Non-Proliferation: Cause for Concern*, PPNN Occasional Paper 2, (Southampton: Centre for International Policy Studies, Univ. Southampton, 1988); Marie-France Desjardins and Tariq Rauf, *Opening Pandora's Box? Nuclear-Powered Submarines and the Spread of Nuclear Weapons*, Aurora Papers 8, (Canadian Centre for Arms Control and Disarmament, 1988); Dan G. Haglund, *The Canadian SSN Program and the Nonproliferation Question*, Centre for International Relations Occasional Paper 29, (Kingston, Ontario: Centre for International Relations, Queen's Univ., 1988)

9 On transit issues Milhollin, op. cit., and Deutscher Bundestag, 11. Wahlperiode, Drucksache 11/3995. Unterrichtung durch die Bundesregierung. Bericht der Bundesregierung an den Deutschen Bundestag über eine mögliche Beteiligung deutscher Firmen an einer C-Waffen-Produktion in Libyen, Bonn, 1989

10 The well-known "small quantity case" involving heavy water exports from the Soviet Union through Switzerland to India is analysed by Milhollin, op. cit

11 Leonard S. Spector, *The New Nuclear Nations*, op. cit

12 Gary Milhollin, "Heavy Water Cheaters", *Foreign Policy*, Vol. 69, (Winter 1987/88) pp. 100–118, makes a powerful case for holding "illegal" importers responsible

13 See reference 1, pp. 82–89

The Non-Proliferation Regime – Stronger than the NPT

Ove Bring

The Non-Proliferation Treaty

The pledge of the nuclear-weapon states in Article VI of the Non-Proliferation Treaty to pursue comprehensive disarmament negotiations in good faith can scarcely be said to have been fulfilled. Apart from the INF (Intermediate Nuclear Forces) Treaty, the only result that has been achieved since the signing of NPT in 1968 – the Bacteriological (Biological) Weapons Convention of 1972 – does not address the problems of the arms race spiral. The arguable lack of fulfillment on the part of the nuclear weapon states takes on a special dimension when related to the criticism of NPT for being unbalanced in its distribution of rights and obligations (i. e. discriminating against non-nuclear-weapon states, especially the developing states). In addition to this criticism relating to discrimination and inequality, the Treaty has also come under attack because it is doubtful whether it can fulfill its purpose. The parts of the NPT aiming to promote the development of civil nuclear energy (Articles IV–V) may thus come into conflict with the parts of the Treaty on non-proliferation (Articles I–III). Furthermore, a great danger lies in the fact that non-nuclear-weapon states which have chosen to remain outside the NPT can develop their civil nuclear energy programmes without the control of the IAEA. Countries such as Argentina, Brazil, India, Israel, Pakistan and South Africa (all of which are outside the Treaty) may thus acquire or maintain a nuclear option. Should one or several threshold states "go nuclear", there is a considerable risk that certain Parties to the NPT will terminate their commitments under the Treaty. The NPT system may then face collapse, particularly since the nuclear-weapon states of France and China are already outside of it today.

At the same time, there is reason to underscore that the NPT has succeeded up to now in its aim of preventing proliferation. After all, an impressive number of countries have acceded to it. According to Article IX:3 of the Treaty, it would enter into force when it had been ratified by the three Depositary States and 40 other states. This stipulation was met in March 1970. Today the NPT has been ratified by 139 states. One of the Treaty's strengths is the fact that it expresses the view of an overwhelming majority of the world community that nuclear proliferation is a serious threat to international security and a phenomenon that is incompatible, or should be incompatible, with public international law as such. As yet, no contracting Party has withdrawn from the Treaty – or breached it.

The Non-Proliferation Regime

Looking to the future, a storm threatens in that, under Article X:2, a conference shall be convened twenty-five years after the entry into force of the Treaty, that is to say in the year 1995, to decide whether "the Treaty shall continue in force indefinitely, or shall be extended for an additional fixed period or periods". This decision is to be taken by a majority of the Parties to the Treaty. We will then see whether the NPT system can withstand the tensions that have accumulated over the years. It is to be hoped that the situation will have improved much earlier through the superpowers' participation in a Chemical Weapons Convention, a comprehensive test ban, or in some other agreement in the field of disarmament. On the bilateral level the INF Treaty is less important as an arms reduction commitment, than as a development in the field of verification. Although the START discussions indicate that further quantitative reductions may come about with regard to the nuclear arsenals, they are not enough. The nuclear arms race could in fact continue through qualitative improvements. It is in this perspective that the need for a comprehensive nuclear test ban should be viewed.

Should the threats against the NPT be averted, more tangible results of arms limitation will be needed. Although such developments are needed anyway (irrespective of the fate of the NPT), one point now to be made is that the regime of non-proliferation is *more than a treaty regime*. Withdrawal of states party to the NPT, motivated by a non-party's possession of nuclear weapons or by non-performance of the treaty obligations on the part of other parties (see Article 60 of the Vienna Convention on the Law of Treaties) [1], may be destructive to the treaty regime, but is cannot change the legal status of that part of the non-proliferation regime which is based on customary law.

Customary Law

Contemporary international law is influenced by a non-formalistic trend of flexibility with regard to the sources of law; recognizing as "law" those rules that meet the moral and legal standards of the contemporary, pluralistic world community. As Professor Edward McWhinney has pointed out, international law is becoming a dynamic process of (*inter alia*) "creative adjustment of old legal doctrine to new societal conditions and demands in the World Community" [2].

Traditional international law doctrine would classify as customary law only those rules that: (1) conform to a general practice of states (an objectively ascertainable normative pattern, *usus*); and (2) express an opinion on the part of states that the norms in question have the status of law (an *opinio juris*). A practice of states common for a certain geographical region could be classified as regional or special customary law. The notion of special custom is not, however, limited to geographical areas. Any grouping of states with common values may establish legal norms applicable between them.

The trend today shows a measure of flexibility in the sense that a societal consensus will be looked for in either a general *usus* or a general *opinio juris* (and not necessarily in a combination of the two), and that this normative consensus could be established through a variety of sources, e. g. the resolutions of international conferences or other meetings between states. The emphasis today has shifted from reliance on closed, formal categories of sources to a more nuanced fact-oriented approach, in which the enquiry, as to a claimed principle of law, "is directed to whether the parties involved, expressly or by their conduct, regard the proposition concerned as normative and legally binding upon them" [3].

At a future-oriented international legal seminar in Sweden in 1981 it was stressed that a broad notion of state practice is called for, so that any evaluation of the content of the law would include statements *in abstracto* and resolutions by the UN General Assembly or other similar organs as possible constituent elements of the law [4]. "Also with respect to the manifestation of an *opinio juris* a flexible approach seems to be in order. Express statements to the effect that a given rule is or should be legally binding are not always necessary; the *opinio juris* may be inferred from other conduct as well, although it is often manifested by the use of normative (legal) language" [5]. Any governmental statements and manifestations of a normative attitude which can be objectively ascertained are relevant in this law-creating context.

The Non-Proliferation Regime and Customary Law

Some of the basic elements of the NPT regime, namely the "freeze" on the number of nuclear-weapon states, the rule of non-acquisition of nuclear weapons applicable to the great majority of states (Article II) and the rule of non-support to weapons production applicable to all states (which follows from Articles I and II), have been lifted forward as independent norms at the first UN special session devoted to disarmament (1978) [6]. The NPT Review Conference in 1985 adopted by consensus a Final Declaration which included the following text:

"The Conference agreed that the strict observance of the terms of Articles I and II remains central to achieving the shared objectives of preventing under any circumstances the further proliferation of nuclear weapons and preserving the Treaty's vital contribution to peace and security, including the peace and security of non-Parties."

Thus an independent *opinio juris* may have emerged with regard to the norms of non-acquisition (Article II) and non-support (Article I plus II). For those states which have also formally accepted the same norms in becoming parties to the NPT, this more limited, but basic, non-proliferation regime may well have acquired the status of special or particular customary law.

If it could be claimed that the non-proliferation commitment in the NPT has the status of particular customary law irrespective of the NPT, the prohibition against acquiring nuclear weapons, as well as the prohibition to help others "go nuclear", would be absolute and binding for states which have once acceded to the NPT. This is of particular relevance with regard to the NPT's future existence. A non-nuclear-weapon state choosing to withdraw from the NPT regime (for instance in order to protest against the superpowers' failure to fulfill Article VI), would still be under obligation to forgo nuclear weapons of its own. The fundamental rules in the NPT would thus have begun to have a life of their own outside the Treaty regime. It is important to keep this line of argument in mind in the face of the wave of mass withdrawals which may threaten the NPT in 1995.

References

1 Article 60 of the Vienna Convention of 1969 has *inter alia* the following wording:

"A material breach of a multilateral treaty by one of the parties entitles:

...

(c) any party other than the defaulting State to invoke the breach as a ground for suspending the operation of the treaty in whole or in part with respect to itself if the treaty is of such a character that a material breach of its provisions by one party radically changes the position of every party with respect to the further performance of its obligations under the treaty."

2 Edward McWhinney, *The International Court of Justice and the Western Tradition of International Law*, (Dordrecht: Martinus Nijhoff, 1987), p. 21

3 *Ibid.*, p. 31

4 Allan Rosas, Customary Law: From "Universal" in a European System to "Regional" in a World system, *The Spirit of Uppsala, Proceedings of the Joint UNITAR-Uppsala University Seminar on International Law and Organization for a New World Order (JUS 81)*, (Berlin: Walter de Gruyter, 1984), p. 223

5 *Ibid.*, p. 224

6 Paragraph 36 in the Declaration of the Final Document of 1978 declares that "Non-proliferation of nuclear weapons is a matter of universal concern", not thus only a matter of concern to states party to the NPT. In the Programme of Action it was stated (paragraph 65) that:

"The goal of nuclear non-proliferation is on the one hand to prevent the emergence of any additional nuclear-weapon States besides the existing five nuclear-weapon States, and on the other progressively to reduce and eventually to eliminate nuclear weapons altogether. This involves obligations and responsibilities on the part of both nuclear-weapon States and non-nuclear-weapon States ... all States undertaking to prevent the spread of nuclear weapons"

The Final Document was adopted by consensus.

Part II

Nuclear Proliferation: Political Priorities

Drawing the Threshold States into a Regime of Restraint, by Joining the NPT or Otherwise [1]

David Fischer

The "Threshold States": A Diverse Group

The "Threshold States" are held to be India and Pakistan, Israel, South Africa, Argentina and Brazil [2]. No single measure or group of measures would suffice to bring all six into the non-proliferation regime. What each of the first four have in common is a perceived threat to national security from without or within (or both). This perception has been one of the main reasons each has kept the nuclear weapon option open. At this point the common factor ends, each perceives a different threat or combination or intensity of the threats. Beyond national security there are other motives that have helped to lead nations towards and beyond the threshold such as considerations of national pride and prestige (e. g. being at the same table as the five permanent members of the UN Security Council) and of the ability to project power [3].

It is obvious that if we wish a nation to withdraw from the threshold the first task is to address the causes that impelled that state towards it. In the cases of India, Pakistan, Israel and, to a lesser degree, of South Africa, this would imply that steps towards a regime of nuclear restraint can only come if they are accompanied by progress in settling issues for which the world has found few if any answers during the last forty years – but over which a few uncertain gleams of hope are beginning to hover.

It has also proved impossible, at least until now, to persuade any state to renounce nuclear weapons once it has acquired them [4]. It is urgent to seek and devise even partial constraints on proliferation before additional states irreversibly cross the threshold.

Argentina and Brazil

This is the most hopeful situation today. In the 1970s Argentina and Brazil seemed each to be determined to be the first Latin American nation to demonstrate her ability to carry out a nuclear test [5]. The advent of democratic, civilian governments has already brought a turn-around in nuclear relations between Argentina and Brazil. Cooperation has at least partly replaced competition; the two heads of state have visited each other's most sensitive plants; they have set up several joint committees and have arranged periodic confidence-building visits. Each is providing the other with some components for its nuclear programme, and these programmes have themselves been severely curtailed. The military ambitions of both countries seem today to be directed towards nuclear submarines rather than nuclear weapons; perhaps they always were.

Democracy is still fragile in both countries. Moreover, each has begun to produce unsafeguarded fissile material. Full acceptance of the Tlatelolco Treaty would avert any danger of a renewal of their earlier nuclear competition.

India and Pakistan

India rejected the NPT in the late 1960s as a discriminatory treaty that perpetuated the right of those that already had nuclear weapons to go on testing and accumulating them and denied them only to the nuclear have-nots. In 1988 India proposed an "alternative" Treaty to UNSSOD 3 (which proposal, if embraced by the NNWS, would almost certainly end up by leaving the world without any Treaty at all). Pakistan professes to be ready to sign any non-proliferation treaty or comprehensive safeguards agreement, international, regional or bilateral, as long as India does the same, at which point India professes to be ready to join a NWFZ provided China also joins.

India has a large and growing stock of unsafeguarded plutonium. She showed in 1974 that she had mastered the technology of making nuclear explosives. Pakistan has a small and growing stock of unsafeguarded enriched uranium, (probably highly enriched) and she could assemble a nuclear explosive/weapon at short notice (if she has not already done so, as some believe).

Although there are considerable constraints against either nation openly declaring or demonstrating that she is a nuclear weapon state, there is an obvious danger that one of the pair (more probably Pakistan which has the greater military incentive to do so) may take a step that would precipitate a nuclear arms/missile race, a race that would burden their economies with vast and spiralling costs and expose them to the risk of nuclear war.

Until recently the outlook seemed bleak. The end of the Afghan war, the improvement of Indian and Soviet relations with China, the first steps towards peace in Indo-China, the advent of a democratic government in Pakistan, the meeting at the end of December between Benazir Bhutto and Rajiv Gandhi and their signing of an agreement prohibiting attacks on the other's nuclear installations [6], and, most surprisingly, the report that India will sponsor Pakistan's re-entry into the Commonwealth, all tend to reduce the underlying tensions that are at the root of Indo-Pakistani nuclear rivalry and of India's fears of China [7]. It is essential to take advantage of the momentum that detente in South Asia may generate.

India and Pakistan is each heavily dependent upon a superpower for economic as well as military support. The US and the USSR should place non-proliferation high on the agenda in their relations with the sub-continent. This would imply alleviating concerns about superpower intervention in the Indian Ocean and Central Asia, shifting the focus to economic rather than military aid (no nuclear submarines!) attaching stricter conditions to military aid and sharing superpower experience in verification and other confidence-building measures.

The consortia that channel and coordinate economic aid to the two countries can help by giving priority to energy investments that increase interdependence, e. g. a subcontinental electric network or grid covering India, Pakistan, Bangladesh and possibly Nepal, and tapping the vast hydro-electric potential of the northern mountain ranges. The consortia should also make it clear that they are not in business to finance a nuclear arms race.

It is most unlikely that either India or Pakistan will formally join the NPT, but this should not preclude some form of participation in the non-proliferation regime. Neutral and non-aligned countries such as Sweden and Mexico whose Prime Ministers are, like Rajiv Gandhi, members of the five continent group, could play a helpful role by explaining the reasons why they decided to renounce nuclear weapons despite the proximity of a superpower. Spain as a recent adherent to the NPT could also explain why, after much hesitation, she decided to join.

There are opportunities for cooperation in the nuclear-electric field. Both India and Pakistan rely chiefly on Canadian designed HWRs in their nuclear power programmes [8]. After 1974 Canada terminated all assistance including assistance in nuclear safety matters. Pakistan has frequently complained about this cut-off. The two countries could pool their extensive experience so as to ensure the safe operation of their HWRs. Perhaps in such circumstances Canada might be prepared to renew cooperation in nuclear safety matters, and the IAEA could play a catalytic role.

A more ambitious venture would be to foster Indo-Pakistani cooperation in the manufacture of HWRs thus capitalizing on their experience in this field, enlarging the market for such plants, enabling Pakistan to expand her nuclear power programme which is now blocked because of her refusal to accept comprehensive safeguards [9], and creating nuclear interdependence in place of nuclear rivalry, possibly with some agreed division of labour as in the case of Argentina and Brazil.

If the relationship further improved one might envisage additional confidence-building measures such as

- visits by the two leaders to each others sensitive facilities (as in the Argentina-Brazil case),
- a mutual renunciation of all nuclear tests including PNEs,
- a mutual inspection regime.

To ensure the continuation of the $ 4 billion American aid programme to Pakistan the President must certify to Congress each year that Pakistan has not manufactured a nuclear weapon. There are reports that the previous Administration recommended against any further certification. There could hardly be a more inopportune moment to discontinue the aid programme – just as Indo-Pakistani relations are improving and Pakistan has acquired a democratic government.

South Africa

South Africa's nuclear capability derives from her position as a major producer of uranium (4000–6000 tonnes per year) and the enrichment technology which her scientists developed in the 1960s but which was translated into an industrial process only with help from Western Europe (especially industrial firms and laboratories in the FRG) and from the USA. A pilot enrichment plant (10 tonnes SWUs per year) has been in operation since 1977 and a "semi-commercial" plant (300 tonnes SWUs per year) came into operation in 1988. Neither plant is under IAEA safeguards.

In 1987, faced by a recommendation by the IAEA Board of Governors that South Africa's rights and privileges as a member of the IAEA should be suspended, President P. W. Botha announced that South Africa hoped "soon [to] be able to sign the NPT". In August 1988 the South African Foreign Minister and the Minister of Economic Affairs and Technology held meetings in Vienna with representatives of the three NPT depositary governments (USSR, UK, USA) and subsequently with those of some 30 other countries. The South Africans also circulated a statement outlining their concerns (chiefly whether, if they joined the NPT, they would be treated like other parties, be able freely to market their uranium and have equal access to nuclear technology, and be able to take part in IAEA General Conferences and NPT Review Conferences). The three depositary governments issued a statement that they were "... resolved to press South Africa further to accede to the NPT."

The General Conference's decision about South Africa has been delayed until September 1989; if by then she has not acceded it is likely that her membership of the IAEA will be suspended. What direction the new Prime Minister (de Klerk) will take is still unclear. South Africa is the only threshold state that might be persuaded to adhere to the NPT by a combination of direct pressures and certain reassurances. The reasons lie in South Africa's particular strategic, political and economic circumstances.

Unlike Israel, India and Pakistan, South Africa is under no military threat from abroad; on the contrary South Africa has projected an armed threat into the rest of Southern Africa. The withdrawal of Cuban troops from Angola removes the only other force that conceivably could have challenged South Africa's military predominance. Nuclear weapons are essentially irrelevant to the internal forces that threaten the South African regime and with which it must eventually come to terms or succumb.

Uranium represents a small but significant proportion of South Africa's exports [10]. The USA has embargoed imports of South African uranium; the UK ceased importing from Southern Africa several years ago; the Japanese government has directed Japanese utilities not to renew contracts for purchases from South Africa when they expire; Taiwan is looking to Australia and the USA for supplies, and the door to Spanish utilities is also being closed. The only significant markets that are not (at least yet) being closed are France and the FRG.

At the same time the independence of Namibia will make that country a respectable source of supply and add to the list of uranium exporters that will compete with and may displace South Africa. It is not surprising that recently the head of South Africa's uranium marketing organization, NUFCOR, announced that there would be a 50 per cent cut in production in the 1990s. A final rejection of the NPT and suspension from the IAEA might further reduce South Africa's attractiveness and perceived reliability as a supplier of uranium.

South Africa has two 965 MW French supplied nuclear power reactors and has shown some interest in importing a third during the 1990s. No state will be willing to supply South Africa unless it has acceded to the NPT.

In short, South Africa has some economic incentives to join the NPT and no significant security incentives to refrain from doing so. There might also be some political advantage in joining. South Africa would be in the (for her) unusual position of being a member of a club in which most of the other parties would wish her to stay. Joining the NPT might also encourage her neighbours, Mozambique, Zimbabwe. Zambia and Angola to adhere and open the prospect of a NWFZ in Africa (or, at least, South of the Sahara) of which South Africa would be an important party – thus marginally diminishing her isolation. It might also open the prospect of better relations with Eastern Europe.

The dilemma that faces many other states is that extending any assurances to South Africa about the security of her uranium markets and her access to nuclear technology, as well as about her rights to take part in the IAEA's General Conferences and NPT Review conferences, would be at odds with the policies that many governments follow with a view to eliminating apartheid. The issue is chiefly of concern today to African states and to the Common Market countries to whom South Africa sells her uranium and from whom she would probably seek to purchase additional nuclear power plants. But the three NPT depositary governments, including the USSR, can influence South Africa's decision.

Israel

It is widely held that Israel is the only threshold state that already possesses a nuclear arsenal [11]. Israel probably crossed the threshold in the mid- or late 1960s. Many Israelis are convinced that their nation's survival is under constant threat. Whether they are right or wrong, persuading Israel to recross the threshold and forego her nuclear arsenal might be as difficult as persuading Britain or France to give up their own nuclear forces. Any movement of Israel towards a regime of nuclear restraint is thus contingent upon progress in settling apparently insuperable political problems.

Until recently the prospects for such movement seemed to be getting bleaker. The Intifada, the fact that Israel's neighbours are acquiring medium-range missiles as well as the means to produce chemical warfare agents seemed likely to increase Israeli reliance on a nuclear deterrent – e. g. against the threat of a militarily paralysing first strike with chemical weapons.

The latest developments have given some grounds for hope. They include the PLO's unequivocal recognition of the state of Israel, the PLO's renunciation of terrorism, Israel's improving relations with Moscow and Eastern Europe, an apparent diminution of Israeli resistance to an international conference on the Middle East and to negotiations involving the PLO and a more pragmatic approach by Washington to the problems of the Middle East.

This may be only another false dawn of which there have been so many since 1948. But in all circumstances it is vital that the USA should continue to persuade Israel to refrain from openly proclaiming her nuclear capability. An Israeli nuclear test or public declaration that she was an NWS would send a destabilizing shock-wave throughout the Middle East which could also have damaging consequences for the 1990 and 1995 conferences and for the future of the NPT. It would face the USA with a painful dilemma. If the USA failed to react with sufficient vigour, her own non-proliferation policy would suffer a severe loss of coherence and credibility.

More generally the superpowers and the EC should place non-proliferation higher on their agenda for a Middle East settlement. Concern about the spread of chemical warfare capability and the role of Western suppliers should not divert attention from the greater long-term risk of nuclear proliferation.

The use of the DIMONA reactor to produce plutonium for Israel's nuclear arsenal, the Israeli bombing of the TAMMUZ reactor in Iraq, and the Iraqi bombing of the half-completed LWR power plant [12] near Bushehr in Iran, surely makes it obvious that the supply of a significant nuclear facility to the Middle East brings a nuclear factor into existing tensions, may provoke armed attack [13], and eventually lead to the spread of chemical warfare capabilities to counter the nuclear force of the enemy.

At present the prospects for a Middle East nuclear-weapon-free zone are poor. It is, nevertheless, desirable to continue exploring Egyptian and Israeli proposals for such a zone. It might be possible to identify the main elements of a nuclear-weapon-free regime that would be acceptable if the political environment became more propitious. As Churchill said, jaw-jaw is better than war-war!

Conclusions

In the last ten years each of the threshold states has acquired or expanded her technological ability to manufacture nuclear weapons. The decision to cross the threshold is, however, political and the likelihood of such a decision has diminished in Argentina and Brazil and probably in South Africa. Moreover there are no other threshold countries in the pipeline and probably will not be before 1995 or later.

The real and pressing danger of a nuclear arms race is in South Asia. Nevertheless, several constraints already operate on Pakistan (the fear of jeopardizing $4 billion of US aid, provoking the USSR and Pakistan's neighbours, and of precipitating a race with India that Pakistan would be likely to lose). There are also constraints on India (loss of aid, provoking a Chinese response, the loss of the moral high ground, and little incentive to take an irreversible step unless Pakistan forces her to do so). Above all, Pakistan and India must weigh the costs and risks of a nuclear arms race and of preparing for and even of fighting a nuclear war. Each must look over the brink and into the abyss as each of the nuclear weapon states has had to do.

In this case, and indeed in each of the others, there are also some hopeful signs. It is essential that the USA and the USSR, the Western industrial world, and the neutral and non-aligned countries should take advantage of the current "wave of peace" to reinforce and supplement the constraints that already exist.

References

1 Besides SIPRI, at least four non-governmental groups are actively promoting international cooperation to discourage the further spread of nuclear weapons:

- The Programme for the Promotion of Nuclear Non-Proliferation (PPNN) co-directed by Benjamin Sanders, formerly director in the UN Department of Disarmament, and Dr. John Simpson of Southampton University. At present this multinational group focuses specifically on the preparations for the 1990 NPT Review Conference;
- The Peace Research Institute of Frankfurt (PRIF) whose non-proliferation work (chiefly directed to Western Europe) is under the direction of Dr. Harald Mueller;
- The Centre for European Policy Studies (CEPS) in Brussels, also chiefly directed to Western Europe; its senior researcher is Dr. Peter Lomas;
- "Non-Proliferation Initiatives" in Washington D. C. whose Executive Director is Dr. Rodney Jones.

PRIF, CEPS, "Non-Proliferation Initiatives", and the Norwegian Institute for International Affairs arranged a conference in Oslo in May 1988 (under the chairmanship of Dr. Johan Holst, the Norwegian Minister for Defence) which specifically addressed the proliferation problems of the Middle East and South Asia. Several of the recommendations in this paper emerged from that conference

2 None has joined the NPT; each has unsafeguarded nuclear plants capable of making nuclear weapon material. Chile and Cuba have nuclear reactors and have not joined the NPT, but all their nuclear installations are at present under IAEA safeguards. Questions have been raised about the long-term intentions of both Koreas and of countries like Libya and Iraq, but they are all parties to the NPT, and amongst them, only South Korea has a significant nuclear programme today

3 There have been numerous studies of national motivations and capabilities, e. g., J. Goldblat (ed.), *Non-Proliferation: The Why and the Wherefore*, (London: SIPRI, Taylor and Francis, 1985) and Mitchell Reiss, *Without the Bomb: The Politics of Non-Proliferation*, (Colombia, 1987) as well as the annual surveys of proliferation dynamics by Leonard Spector of the Carnegie Endowment for International Peace

4 Only after forty years of nuclear competition were the superpowers able to take the first step to reverse the nuclear arms race. There are indications that it may prove even more difficult to persuade second-rank nuclear powers to follow suit

5 One does not know whether this was really the intention of the governments concerned or whether the search for nuclear autarchy should rather be seen as a defiant gesture. There was a good deal of resentment in both countries about the "privileged" position of the nuclear weapon states and at the controls the latter sought to impose on others, but were not prepared to accept on themselves

6 Under the agreement each nation will have to give the other a complete list of her nuclear installations – a step away from suspicion breeding secretiveness (*International Herald Tribune*, 1 February 1989)

7 The recent announcement (*International Herald Tribune*, 5 February 1989) that Pakistan had successfully launched a short range (300 km) surface-to-surface missile with a payload of 500 kg and Mr. Rajiv Gandhi's somewhat threatening reaction cautions against too much optimism

8 In the 1960s India also imported two LWRS from the USA. Recently India concluded an agreement with the USSR to supply her with two VVER-type LWRs of 1000 MW each; however, the Indian energy authorities made it clear that they will continue building HWRs. Pakistan has been on the market for several years for a 600 MW PWR, but no exporting state has been prepared to submit tenders

9 If such cooperation ever did take place India would have to face some difficult problems of safeguards. She could not insist on full-scope safeguards in Pakistan (in return for the acquisition of nuclear power plants) unless she were prepared to accept such safeguards herself. On the other hand she would presumably be unwilling to agree to any arrangement which would enable Pakistan to acquire an unsafeguarded source of plutonium. The obvious solution would be an agreement under which the IAEA would safeguard the reactors themselves, their fuel and its products and any replicas of the reactors

10 The exact value of South Africa's uranium exports is secret, but it is probably of the order of 3 % of all South african exports. However, the profitability of certain gold mines is related to the amount of uranium that they co-produce with gold, and the loss of a market for their uranium might lead to a drop in gold exports

11 By the Pokharan test in 1974 India showed that she had mastered nuclear explosive technology, but Indian governments have since consistently maintained that India has not become a NWS. India has a large and growing stock of unsafeguarded plutonium, but does not appear to have undertaken a nuclear weapon programme.

Pakistan may be one or more "turns of a screw driver" from completing the manufacture of a nuclear explosive device, but in 1988 the US government still felt able to certify that she had not taken the final step.

South Africa was apparently preparing for a nuclear test in the Kalahari desert in 1977, but subsequent developments have raised doubt whether she had enough HEU for a test at that time or even in 1979 when a US Vela satellite picked up a double flash over the South Atlantic Ocean – an event that is usually characteristic of a nuclear test. A panel set up by President Carter concluded that the flash probably did not indicate a nuclear test. There are other circumstantial reasons for doubting whether South Africa has crossed the threshold,

though little doubt that she would be able to do so at short notice – as her Foreign Minister declared in a statement in Vienna last year.
However close to making a nuclear explosive device India, Pakistan and South Africa may be, there seems little doubt that they are all well behind Israel in the field of missile technology and in their ability to equip their aircraft, missiles or long-range artillery with nuclear warheads

12 So far only the civil engineering work has been undertaken at the Bushehr plant

13 There are reports that since the Iran/Iraq armistice a German/Argentinian company has undertaken to complete the Bushehr plant

Towards a Universal Framework of Nuclear Restraint

Munir Ahmad Khan

Recently there has been a perceptible relaxation in global and regional tensions all over the world. The most noticeable sign of this "thaw" has been in the relations between the two superpowers. The leaders of the USA and USSR have publicly shared the vision of a nuclear-weapon-free globe, and the first tentative steps in the direction of nuclear disarmament have been taken. The general atmosphere is thus more conducive than ever before to evolving a framework of nuclear restraint which ought to be nondiscriminatory in principle, universal in application and effective in implementation.

Rationale for Nuclear Restraint

There are many imperatives for pursuing a policy of nuclear restraint. First and foremost is the moral argument which applies to nuclear arms as well as to other weapons of mass destruction like chemical and biological weapons. No matter how well and specifically targetted, they inevitably involve non-combatants and spread destruction indiscriminately over a large area beyond the frontiers of the belligerents. These weapons have, therefore, always aroused feelings of horror and revulsion among all humane people of the world.

Strengthening the moral argument is the more pragmatic reason that nuclear weapons will now serve no military purpose whatsoever. A nuclear war is inherently unwinnable and nuclear weapons are thereby basically unusable. In a rational world, therefore, there should be no nuclear weapons anywhere. Mankind's hope lies in striving to move towards this rational state of affairs through dismantling existing nuclear weapons and ensuring that no more are made. Both these aspects of the issue must be addressed comprehensively in order to establish an equitable, stable and acceptable framework of nuclear non-proliferation throughout the world.

From an international political perspective, too, it would be unwise for a non-nuclear-weapon state (NNWS) to cross the nuclear threshold overtly by staging a nuclear explosion. There would be no tangible gain for it on the international scene; rather there are bound to be adverse reactions. Similarly, any addition in the nuclear arsenals of nuclear weapon states (NWS) will only enhance their "overkill" capacity and have no strategic value. It is thus in the common interest of all states to exercise restraint in regard to both vertical and horizontal proliferation.

Why Nations Want to Go Nuclear

From a purely technical point of view the nuclear option is available to many countries including the so called threshold ones. However, exercise of this option and its translation into actual nuclear capability, with all the perceived benefits and attendant costs, requires a conscious decision by the nation concerned. It has been perceptively observed by the ANS Special Committee on Nuclear Weapons and Peaceful Uses of Nuclear Energy:

"... interest in nuclear weapons is more likely to arise from complex considerations of national security, rather than from technical opportunity ... once a decision to acquire nuclear weapons is made, the acquisition of sufficient material is only a matter of time, available skills and resources. Nuclear weapon proliferation is thus primarily an international political issue, and the eventual solution is in the political domain rather than in the technical". [1]

Apart from security considerations, a nation may be impelled by domestic or regional compulsions to retain a nuclear option. For some governments, the pursuit of nuclear capability may serve to surmount internal political difficulties. There are other countries which are known to represent the interests of certain advanced countries and have virtually a status of client states. In such a situation, the responsibility for enforcing restraint lies more with the concerned patrons than with the would-be proliferator.

In some cases a nation may chase the nuclear option in its desire for an illusory prestige. There is yet another incentive which applies particularly to a large state wishing to establish political hegemony and economic domination over its smaller neighbours. The nuclear weapon can be regarded as a key card by such a country in its power game.

The nuclear ambitions of Israel and South Africa are in a class by themselves. These states are considered by their neighbours to be extensions of Western power and their culture into their territory. These two states seek security by building a nuclear fortress in order to secure their survival in politically and culturally alien surroundings by dint of sheer military superiority. They have, therefore, resorted to a policy of ambivalence in regard to their nuclear capability. They neither admit nor deny the possession of nuclear arms.

It is thus seen that, while the so-called threshold states share some common ground for retaining their nuclear option, they are mostly a heterogeneous group as regards their capabilities, motives, aspirations and intent. We, therefore, have to address both the common features and the differences in this regard, and understand their specific situation. The future non-proliferation arrangements must be so fair, above board and universal that these states are left with no reasonable excuse to stay out.

Disillusionment with NPT

By its very nature, legitimizing the possession of nuclear weapons by five states makes the Nuclear Non-Proliferation Treaty (NPT) inherently discriminatory. This is why even many signatories question its very moral basis. Only those states which are affluent, have a stake in the status quo and have guaranteed assurances about their security, support this Treaty without reservations. The LDCs in general have a negative perception of the NPT and even those who subscribe to it, feel that they have derived little benefit from it. The so-called threshold states among developing countries are more vocal in their opposition to the NPT. In their view it is aimed at perpetuating the nuclear hegemony of the nuclear weapon states. They also resent the implied accusation of irresponsibility on the Third World in general. They further point out that there are no checks on the NWSs while the NNWSs are effectively required to surrender part of their sovereignty in return for vague and as yet unfulfilled promises of nuclear cooperation. This unequal treatment is clearly unacceptable, and the fears and misgivings of LDCs have hardly been allayed by the subsequent behaviour of major supplier states whose actions seem to be solely dictated by their respective national and ideological interests.

The NPT was supposed to be based on a *quid pro quo* consisting of a balance of obligations and privileges. The NNWS signatories to the NPT agreed to give up their nuclear weapon option in return for the promised fullest possible exchange of equipment, materials and scientific and technological information for the peaceful uses of nuclear energy, including the use of peaceful applications of nuclear explosions under strict international control. The NWSs, for their part, undertook "to pursue negotiations in good faith on effective measures relating to a cessation of the nuclear arms race at an early date and to nuclear disarmament, and on a treaty on general and complete disarmament under strict and effective international control" (Article VI). But almost before the ink of the signatures had dried out, there began substantive reinterpretation of the Treaty to suit the interest of the superpowers and other nuclear advanced countries. The provision calling for nuclear disarmament was relegated to the background, and primary stress was laid instead on controlling the spread of nuclear technology. Although clearly provided for in the Treaty (Article V), benefits of peaceful uses of nuclear explosions have not been generously shared with NNWS signatories while the superpowers continue to exploit PNEs for underground seismic surveys, explorations and engineering. Even the sharing of nuclear fuel cycle technology is done very selectively and is subject to political conditions. As a result, there have evolved categories of membership among the signatories of the NPT: first, there is the "inner circle" comprising a few countries of the OECD; second, there are the "trustworthy" client states; and lastly the rest for whom the NPT "bargain" is an expanding set of obligations matched by a diminishing set of privileges. The ever-changing rules of the game give the impression that, while the nuclear powers would gladly acquiesce in selective proliferation, they would be most reluctant to share peaceful nuclear technology with the needy countries of the Third World.

Why Nuclear Power

Economically, the most logical use of nuclear energy is for generating electric power. This application has been confined almost entirely to advanced countries, there being just 25 units in developing countries totalling around 15,000 MW, that is, less than five per cent of the aggregate nuclear installed capacity of the world. If the present trend continues, then the entire Third World will have no more than 40,000 MW nuclear by the year 2000, that is, less than that in a single European country at present. The oil-importing developing countries have fossil fuel proven reserves of just 11 TOE per capita, versus an average of 151 for the whole world and 366 TOE per capita for developed countries. Further, the ability of LDCs to pay for energy imports is very limited. If they were to consume fossil fuel in the future at the same level as industrialized countries do at present, the carbon dioxide emissions will increase four-fold, leading to extensive pollution and the greenhouse effect. This imposes a serious constraint on their development, yet the economically attractive and environmentally better alternative of nuclear power is denied to them. This creates misgivings and doubts in their mind about the real nature and intent of the non-proliferation regime. There are those who perceive in this order of things a new type of technological colonialism involving transfer of valuable mineral and energy resources from LDCs to the industrialized countries (in 1987 alone, 1500 million tons of oil were produced in the South, of which 800 million tons were exported to the North) without adequate economic compensation or sharing of science and technology for the former's development. As the disparity and awareness of this inequitable treatment increase, dismay sets in, leading to repudiation of the current non-proliferation regime.

Industrialized countries, on their part, base the withholding of nuclear technology on the alleged link between nuclear power and nuclear weapons. It has been variously argued that use of nuclear power is "the main driving force behind proliferation" and that "power reactors can be considered large-scale military production reactors with an electricity by-product rather than benign electricity producers with a militarily unattractive plutonium by-product". Of these, the first assertion does not stand up to the testimony of history; all the five recognized nuclear powers had acquired nuclear weapons before embarking upon the utilization of nuclear heat for electricity generation. The second contention requires somewhat more detailed examination. If a country had nuclear power stations and a reprocessing plant, all under safeguards, it could in principle reprocess the spent fuel from the reactor to obtain reactor grade plutonium. Such plutonium is not easily usable for weapons and its attempted diversion from safeguards will lead to serious international reaction. Therefore it would be technically impractical and politically unwise for a country to follow the nuclear power route to proliferation.

It is thus clear that the "link" between nuclear power plants (under IAEA safeguards) and nuclear weapons proliferation is highly tenuous. It is unfortunate that certain groups have used this alleged link to create a certain confusion in the minds of laymen and policy-makers and thereby discredit the use of nuclear power. Of course, for the informed leaders – both technical and non-technical – there has never been any confusion. Sir John Hill, former Chairman of UKAEA, has stated:

"Stopping nuclear power programmes will not make the proliferation problem go away: the problem is primarily a political one, and the least that could be said of the proposed strategy of cutting back on nuclear power is that it would prove totally irrelevant to the task at hand . . . ." [2]

The Limitations of Embargoes

Even though there is no real linkage between proliferation of nuclear weapons and development of nuclear power under safeguards, the industrialized countries have chosen to use the denial of nuclear power as an instrument for enforcing the non-proliferation regime. The various manifestations of the policy are well-known: unilateral abrogation of international agreements, violation of Article IV of the NPT for the signatory recipient states, the Nuclear Nonproliferation Act of the United States, and the London Club guidelines. In this whole exercise, the technical and political realities of the 1970s and 1980s, as well as historical lessons, are ignored.

The policy of denial has not been very effective in controlling the spread of any modern technology. Embargoes may serve some economic or political purpose in the short term and introduce delays but, in the long run, will prove to be counterproductive by providing the political motivation to seek nuclear autarky in sensitive areas. This lesson, however, does not seem to have been appreciated even by those countries of the North which have themselves been victims of such denials at one time or another. The reluctance of industrialized countries to share advanced technologies has unwittingly led to the emergence of new supplier states.

France was denied access to nuclear technology in the 1950s. It has not only attained self-sufficiency but is now a major potential supplier of sophisticated nuclear plants and services. Japan has developed a full range of nuclear fuel cycle and power reactor production capabilities and could become a major supplier in the international nuclear market whenever it chooses to do so. In the Third World, Argentina, Brazil, China, India and Pakistan are prospective suppliers to varying degrees. Argentina is already exporting research reactors, and China has also exported nuclear materials and equipment under safeguards. None has yet developed full capability for nuclear power plant export, but if the policy of denial continues on the part of major existing suppliers, there will be strong political and commercial incentives for new supplier states to emerge to fill the vacuum.

Need for Fresh Perspectives and Bold Initiatives

Fifty years after the discovery of fission, nuclear knowhow and technology has already spread beyond recall. We must accept the fact that it is too late to rely on technical fixes which would no longer work. Nuclear power is interdisciplinary and embraces a broad range of conventional and dual-purpose technologies. Unless all technical and scientific knowledge in a multitude of disciplines is restricted it would be difficult to stop the spread of nuclear technology to those who seek it. In fact we have been fortunate that nuclear weapons have so far been confined to only five states, with nuclear explosive capability existing only in a handful more. In view of the accelerating march of science and technology, the proliferation of nuclear weapons knowhow cannot be checked by the outmoded policies of embargoes and restrictions.

There is a need for a fresh perspective on the issue with a view to evolving a framework of restraint acceptable to all, including the so-called threshold states. Any such framework must start off by recognizing the reality that the principal proliferators are the two superpowers themselves. The stockpiles of nuclear warheads available to the USA and USSR are, even if taken separately, sufficient to ensure global annihilation several times over. Their combined nuclear arsenals have the equivalent of four tons of TNT for each inhabitant on Earth. The primary responsibility for nuclear restraint must, therefore, lie with the superpowers who have to set the moral tone for other nations to follow.

Global Nuclear Disarmament

The first and most important element of a viable framework of restraint must be serious negotiations between the superpowers aimed at reducing their nuclear arsenals and moving gradually but firmly towards total nuclear disarmament. It must be recognized that their nuclear arms race has caused wasteful diversion of human and material resources on an unprecedented scale. It is high time that at least a part of these resources are freed and made available for the development of the underprivileged four-fifths of humanity.

Steps Towards Restraint by Threshold States

Threshold states include those NPT nonsignatories which have acquired nuclear capability but have not overtly exercised the nuclear option. Then there may be other states which belong to the NPT but may have incentives to go nuclear in the future. In order to persuade these diverse states to exercise restraint, the following measures may be considered.

Security Guarantees

The NWSs should attempt to understand the motivation for some developing countries to retain their theoretical nuclear option. As discussed earlier, the basic driving forces behind a nation's quest for nuclear weapons are its perceptions of security and national interest as well as a sense of national pride; and we must appreciate that nation's own point of view if we are to take any effective steps to mitigate its concerns. Clearly the smaller states of the world, particularly in regions where they are overshadowed by one or two regional powers, would have the greatest reason to feel insecure. Unless these legitimate security concerns of threshold states are met and dealt with effectively, the political and psychological incentives for them to retain the nuclear option would remain. One effective way to achieve this objective would be for the NWSs to offer dependable guarantees against nuclear attack or blackmail to those NNWSs which abjure the acquisition or use of nuclear weapons through a solemn international declaration or treaty. This could be an important component of an enlightened policy aimed at inducing threshold countries to exercise nuclear restraint.

Sharing of Technology

It can hardly be over-emphasized that the proliferation of nuclear weapons is essentially a political problem which should be tackled primarily on the political plane rather than through the self-defeating policy of trying to use technical fixes. Factually, many NNWSs have at present more of a nuclear industrial base than the US had in the early forties; and, psychologically, it is not acceptable that the right to engage in so-called sensitive technologies should be the prerogative of a few countries considered to be politically safe by the nuclear suppliers group. A LDC, like any other sovereign state, wants to be free to exercise its right to develop, under appropriate safeguards, nuclear fuel cycle programmes based upon its own needs and requirements. The overemphasis on restricting transfer of technology has actually been counterproductive because it has generated a feeling of insecurity among the "have-nots" who do not wish to remain for ever dependent on others for fuel cycle services. This feeling has led to the indigenous development of fuel cycle facilities in a number of recipient states, outside safeguards. Another important element of a stable policy of non-proliferation must, therefore, be to facilitate, not hinder, the transfer of technology of nuclear power generation – under international safeguards – to energy-deficient developing countries.

The two measures suggested above viz., provision of adequate guarantees against nuclear attack and the promotion of peaceful nuclear power technology, were presented during the negotiations leading to the NPT, but security guarantees were not agreed to and the transfer of peaceful nuclear technology under the NPT has not been implemented. It is, therefore, necessary that, whatever measures are agreed upon for the twin issues of nuclear power development and nuclear weapons non-proliferation, an international agency be entrusted with the task of overseeing the satisfactory implementation of international commitments and contracts, and reporting instances of non-compliance. The IAEA, whose safeguards system has functioned reliably and adequately, is such an Agency with the requisite experience in the delicate task of monitoring the implementation of international safeguards agreements. The IAEA would of course need to be suitably strengthened to perform any additional function entrusted to it.

Nuclear-Weapon-Free Zones

Along with the steps towards nuclear disarmament by all the nuclear weapon states (led by the USA and USSR), the problem of horizontal proliferation has to be addressed. The developing world shares the concern of industrialized countries over the possible emergence, overt or covert, of new nuclear states, but the issue should be viewed in the proper perspective. Developing countries, in general, do not possess the economic and technological means or the necessary trained manpower to engage in any significant nuclear weapons programme. Even for so-called threshold countries in the Third World, such a programme would pre-empt a major fraction of available resources sorely needed for their development. Since the LDCs cannot attain a credible nuclear deterrent, possessing a few nuclear weapons will not add to their security and may in fact endanger it. Horizontal proliferation in this sense could indeed cause serious security problems, particularly in a regional context, and lead to nuclear competition between rival neighbouring states in sensitive regions and have global repercussions. This led to several proposals for the denuclearization of different regions and the actual establishment of nuclear-weapon-free zones in Latin America and the South Pacific. These would serve as valuable building blocks for an eventual all-encompassing non-proliferation regime. The continued success of this approach requires full backing of the NNWSs in the regions and the active interest and support of the nuclear powers. This ought to be the third element of an effective approach towards nuclear non-proliferation.

Confidence Building Measures

Also in the regional context, the international community should lend its support to confidence-building measures such as those suggested in the South Asia Region. These include a joint declaration by neighbouring countries to abjure the development or acquisition of nuclear weapons, mutual inspection of nuclear facilities, simultaneous acceptance of IAEA full-scope safeguards, agreement not to attack nuclear installations, and so on. These measures would strike at the very roots of nuclear proliferation, viz., lack of mutual trust and understanding and thus contribute an important restraining influence on all the concerned parties.

Resolution of Regional Political Problems

Since the motivation to go nuclear is essentially political, the desired restraint will also come through the political route. Major regional conflicts and problems may drive certain states to seek weapons of mass destruction including chemical, biological and nuclear armament. This applies particularly to the Middle East, South Africa, South Asia, and Latin America. The nature of the problems is well-known; the need is to find regionally acceptable solutions in which the international community can and should lend a helping hand.

Conclusion

It appears that the NPT as it now exists is neither a necessary nor a sufficient condition for ensuring nuclear non-proliferation. In fact its many deficiencies in conception and implementation make it largely unequal to the task. The essential characteristics of a new enlightened policy against nuclear proliferation would be equity, universality, and reasonable balance between obligations and privileges. The goal of universal nuclear disarmament will be achieved only when all states recognize, on their own volition, that modern science and technology has irrevocably shrunk the globe and that the effects of weapons, of war's destruction, and of environmental pollution cannot be confined to national boundaries. One cannot erect impregnable barriers between nuclear haves and have-nots, rich and poor, North and South. While differences in development may remain and will be acceptable, glaring disparities can only lead to erosion of goodwill, disruption of peace and general instability.

The Chernobyl incident, acid rain and marine pollution are all reminders of the reality that today no nation can afford to be an island unto itself. Whether we like it or not, all five billion of us have been thrown together in an essentially integrated and interdependent world; we have to learn to live safely and harmoniously. That is the challenge of the closing decade of this century; science and technology can help us face the challenge provided we can collectively display the proper perception and political vision for the purpose.

In the final analysis, nuclear non-proliferation is just one dimension of mankind's quest for survival. Both the problem and the solution belong to the domain of international policies – encompassing security, progress and prosperity for all. The North and the South are diverging in their perceptions of an equitable world order. They have to come closer and realize that their survival on this planet must be together and not at the expense of each other.

References

1 *Nuclear News*, August 1983

2 Karl Kaiser, *Reconciling Energy Needs and Nonproliferation*, 1980

Argentina and Non-Proliferation

Julio C. Carasales

Argentina is considered a nuclear threshold country but is not a party to the Non-Proliferation Treaty. This situation explains why questions are frequently raised about the real position of Argentina regarding non-proliferation.

For this reason, it would perhaps be useful to try to describe, as briefly as possible, the principal characteristics of the Argentine attitude on this issue. This is the purpose of this paper.[1]

Argentina and the Non-Proliferation Treaty

Argentina was not a member of the Eighteen Nation Disarmament Committee and, therefore, did not participate in the drafting of the NPT. As a matter of fact, only two states really drafted the Treaty. All the others only expressed their views about the successive drafts, and the two superpowers took those views into account, sometimes, when preparing a new draft. The UN General Assembly considered the Treaty but, as everybody knows, the text coming from Geneva was not really debated and suffered only very minor changes, introduced by the authors themselves – the two superpowers.

At that time Argentina made public several important objections to the Treaty and abstained when the Treaty was finally supported (not "approved") by the Assembly on 12 June 1968.

Since then Argentina has maintained a negative stance toward the NPT. It is highly important to note that this position has been kept by all civilian and military governments of Argentina up to the present time, whatever their different political ideologies. Moreover, the two main political parties of Argentina, those which are likely to govern the country in the future, hold a similar position vis-à-vis the NPT. This clearly shows that Argentina's attitude on this issue is not the product of a particular view or interest but a national attitude, and that this attitude is not likely to be changed in the foreseeable future.

What are the reasons for this position? Some are very well known: the discriminatory nature of the Treaty; the legitimation of the possession of nuclear weapons in the hands of a few states; the obligations, restrictions and sacrifices imposed only on the non-nuclear-weapon states, particularly developing

1 The author wishes to stress that he is expressing personal views, which are not necessarily shared by his Government.

countries, while similar sacrifices are not required from nuclear-weapon states; non-fulfillment by these states of their obligations regarding disarmament and nuclear cooperation; the irrelevance of the NPT on the question of nuclear technology transfers, when severe restrictions and denials are imposed by the supplier states (the 1977 London Suppliers' Guidelines) which go far beyond the obligations assumed in the NPT by the developing countries, precisely in the hope of receiving free and full technical cooperation; the irrelevance of the NPT even as a source of security, considering that a party's nuclear facilities can be destroyed on grounds that the Treaty does not guarantee their peaceful nature; the total prohibition of peaceful nuclear explosions by non-nuclear-weapon states (when those explosions were carried out by at least one nuclear-weapon state, demonstrating that they are practical and economical) while nothing has been done about compliance with the provisions of Article V of the NPT; and so on.

All these reasons – and others not mentioned – make the NPT, at least for Argentina, a highly unsatisfactory Treaty. It could be argued, of course, that, even recognizing the existence of all or most of the above-mentioned defects, the NPT has nevertheless been ratified by the great majority of the international community and that, in any case, the NPT is a source of international security.

Both arguments require some qualifications. It is true that a great number of states are parties to the NPT, but most of them have no significant nuclear development and, therefore, their pledge not to produce nuclear weapons is really symbolic and does not constitute evidence of the merit of the Treaty.

In the case of the developed, non-nuclear-weapon countries party to the NPT, most of them belong to or are closely associated with one or the other of the two powerful military alliances, one of which holds the doctrine of nuclear deterrence, that is, the possibility of using nuclear weapons, while the other, adhering to the position of no-first-use, certainly does not exclude the use of nuclear weapons if necessary. Many of these countries have nuclear weapons on their territories. The real extent and value of the contribution of these countries to a world free of nuclear weapons is, therefore, open to doubt. In Argentina's case, it is also doubtful whether they have the authority – and the same can be said even more forcibly about the nuclear-weapon-states – to preach to a developing country about the need to become a party to the NPT and to accept a set of rules and conditions that would hinder the free development of its peaceful nuclear industry, all for the sake of world peace.

Another line of argument stresses the utmost importance of the NPT as a source of international security. It points out, in support of that assertion, the fact that since the signing of the NPT, no new nuclear weapon state has appeared, at least openly. Conversely, the states which have not ratified the Treaty are presented as being responsible for any insecurity existing in the world. I have personally heard this argument several times from delegates of some nuclear-weapon powers and their allies.

In answer to this, it seems that a state of international security simply does not exist in the world today. And one of the main, if not the main, reasons for such a state is the existence of about 60,000 nuclear weapons distributed all over the earth. It is really ridiculous to attribute responsibility for the aptly named "balance of terror" (which in essence continues to prevail even after recent promising developments) to the relatively few threshold countries which have chosen not to ratify the NPT.

Regarding the contention that the NPT is responsible for the non-appearance of new nuclear-weapon states, the following can be said: most of the developed countries are protected, in some way, by the nuclear umbrella of the superpower to which each is allied, so they do not really need to develop nuclear weapons. In any case, it is safe to assume that the existence of the alliance is the real reason why they have remained or were compelled to remain non-nuclear, not the existence of the NPT.

There are some countries which have sufficient technical knowledge to develop nuclear weapons but which are not members of a military alliance. And these countries are party to the NPT. They have not produced nuclear weapons, but are we justified in attributing such restraint to the prohibitions contained in the Treaty, and not to the sincere political will of the countries concerned, which would no doubt have followed the same policy even if the NPT had not existed? To countries with no nuclear capability the existence of the NPT as the cause of their policy of restraint is a non sequitur.

Finally, the nuclear threshold states not party to the NPT regime have certainly decided not to become nuclear weapon states by a sovereign decision, and it is wrong to attribute the merit of such a policy to a Treaty to which those states are not linked.

Argentina and Non-Proliferation

The preceding raises a pertinent question: if Argentina is not, and is not likely to become, a party to the NPT, what is the country's policy regarding the essential issue: the non-proliferation of nuclear weapons?

First, a clarification is absolutely necessary. Non-proliferation and the Non-Proliferation Treaty are not synonymous, do not mean the same thing. Non-proliferation is a goal, a worthy objective, the NPT is a means to that end, just one means among several possible others, like the Indian proposal in SSOD III. It is totally wrong to conclude, as it is often done, that because a country chooses not to agree to a particular instrument, such a country intends to proliferate or at least to leave all options open.

In any case, this is not the case of Argentina. Time and time again its spokesmen, at all levels of government, including the highest, and before the highest fora, including the UN General Assembly, have repeated the absolutely peaceful character of the Argentine nuclear programme. Argentina began its nuclear activities in the fifties and has developed a fairly advanced nuclear industry. Today it is considered an emerging nuclear supplier which cooperates in this field with several countries in all areas of the world, parties and non-parties to the NPT. For over three decades not a single action, not a single incident, has been made public which could be honestly interpreted as indicating the intention, on the part of any Argentine government, to develop nuclear weapons.

This is a fact. And if nothing has been uncovered certainly it has not been for lack of trying. For instance, the circumstance of two retired naval admirals, who were chairmen of the National Atomic Energy Commission for a number of years, has been mentioned countless times as proof of the military character of the Argentine nuclear programme, totally ignoring the civilian nature of the Commission and its direct subordination to the President of the Republic and not to any branch of the armed forces, or the fact that the two persons involved had technical qualifications which later made them valuable consultants and advisors to the IAEA and to foreign governments.

Perhaps, at one time or another, an individual or fringe organization in Argentina has come out in favour of the acquisition of nuclear weapons. What country does not have its share of irresponsible people? But never has an utterance of such a nature been made by a serious, representative body of opinion. No political party in Argentina favours the development of nuclear weapons, and certainly no member of the Government has ever taken such a position. And this sober attitude prevailed even during the South-Atlantic war of 1982, when more than one Argentinian thought that his country would have been treated differently had Argentina been a nuclear-weapon country.

Argentina took a decision several decades ago to follow an independent road regarding nuclear development. From the very beginning, it established as a goal the acquisition of a complete nuclear fuel cycle. Also, from the start, Argentina opposed any move, intentional or otherwise, to make her a nuclear-dependent country, meaning a country in which the decision about what she could do and what she could not do in nuclear matters would be taken in other places. One could agree or disagree with that decision, but it was a national sovereign decision.

This policy has been consistently pursued by Argentina over the years, very often facing great difficulties, internal and otherwise. It almost always encountered strong international pressure to abandon such a road, always invoking high-sounding reasons that frequently hid motives of a commercial nature and a desire to preserve captive markets, or so it seemed.

The rejection of the NPT by Argentina is to be seen as a position taken with the intention of keeping full freedom of research and development in the nuclear field, without the inhibitions and impediments that the Treaty implies. The technology is neutral, neither good nor bad in itself. What is good or bad is the use to which it is put. To hinder the development or transfer of certain technologies because they could be used to produce nuclear weapons is a grave mistake, when the same technologies could also be used to advance civilian peaceful ends. What is essential is the intention of the government involved. The NPT took the wrong approach, in the opinion of Argentina, and successive Argentine governments decided not to renounce the option of an independent, peaceful nuclear programme, nor to allow it to be subject to the whims or changing policies of other governments.

At the same time, Argentine governments without exception have followed a policy of strong non-proliferation and, as was mentioned before, not a shred of evidence can be shown to the contrary.

Argentina is, as the record proves, at the forefront of all disarmament fora that oppose nuclear weapons. It initiated the UN General Assembly's debate of the item entitled "Prevention of a Nuclear War". Argentina has come out strongly and repeatedly in favour of the prohibition of nuclear weapons tests and of nuclear disarmament. The Argentine Head of State is a member of the Group of Six whose initiatives in the field of disarmament have been widely supported.

Argentina signed the Treaty for the Prohibition of Nuclear Weapons in Latin America on 27 September 1967. It has not ratified it yet, but the reasons are not comparable to those relating to the NPT. The South Atlantic conflict added a cause for concern, and not just to Argentina, because it proved that the commitment undertaken by the nuclear-weapon states in Protocol II not to use or threaten to use nuclear weapons against parties to the Treaty is just a unilateral declaration not subject to any kind of control by the interested countries.

In any case, according to Article 18 of the Vienna Convention on the Law of Treaties, a state which has signed a treaty is bound to respect its purpose and spirit and act accordingly. There is no doubt that Argentina considers itself obliged to respect the main objective of the Tlatelolco Treaty, which is not to develop nuclear weapons in Latin America, and the Argentine observers have so stated on several occasions before OPANAL General Conferences.

A recent development is considered by all to be highly positive: a policy of close cooperation with Brazil, traditionally looked on as a potential rival, in the nuclear field. The transparency of the nuclear programmes of both countries will be a source of confidence and security to each other and, it can be added, to the world.

Finally, it should also be taken into account that Argentina, a new and rapidly growing supplier of nuclear technology and equipment to Third World countries, always requires IAEA safeguards on all of its exports. This is the same policy that Argentina has followed regarding its own imports.

Perhaps this paper could be ended by stressing four points: a) the exclusively peaceful nature of the Argentine nuclear programme, b) the fact that the preceding assertion has not been disproved after more than three decades, c) Argentina's public and strong commitment to the cause of non-proliferation, even if it does not agree with one of its instruments, the NPT, and d) the need to recognize that Argentina deserves to be trusted, because its record does not justify any other attitude, and, conversely, that the criticism addressed to it, however well-meaning, is generally unfair and based on fears and wrong assumptions and not on actions and evidence.

The Non-Proliferation Treaty in the Middle East and Africa

Mahmoud Mahfouz

The global importance of the NPT and its limitations have been extensively debated ever since its original conception and will probably continue to be debated as long as nuclear strategic confrontation persists in any form. In a changing world, the original relevance of the NPT is going to be the determining factor in establishing peace, particularly in regions where political crises and tensions exist. However, there are several reasons for a shift of importance from global to regional parameters:

1. The cooling down of the strategic confrontation between the superpowers offers new opportunities for resolving, or containing, hot confrontations, but it also decreases the willingness of either superpower to become directly involved, or to continue to give open-ended coverage to one side, in a regional crisis.
2. The increasing sophistication of the arms race in many regional conflict areas, particularly the Middle East, Southern Africa, and South-East Asia is alarming, not only because of the disparities, and unfairness of the major suppliers who are fuelling the race, but also because of their growing support for the build-up of dangerous capabilities. Furthermore, the arrival on the arms market of several alternative suppliers is a new disturbing feature of the arms race.
3. In the context of the changing strategic and technological races, differentiation between the various weapons of mass destruction (nuclear, chemical, biological, conventional) is bound to become blurred. The uneven evolution of military capability among conflicting sides, with apparent immunity, or even collusion, will ultimately provoke the acquisition of an achievable retaliatory deterrence by the side that feels threatened.
   The spiral of the strategic arms race enhanced by technology has been transferred to the South, and seems to gather the same catastrophic momentum. Regional conflicts are bound to lead to more fear, suspicion and bitterness.
4. There is also moral degradation and brutalizing, which repeatedly demonstrates the increased unreliability of political as well as military behaviour. In this context, the use of weapons of mass destruction does not appear as impossible as it does to the outside world. This is the perception both of countries who have acquired the weapons, and of those who are threatened.

Threat of Nuclear Proliferation in Africa and the Middle East

The threat of nuclear proliferation in Africa and the Middle East is growing fast, producing political and economic destabilization, as well as promoting a devastating strategic arms race among countries of that region.

The political stability in any region is the major factor which decides the success of any dialogue aimed at disarmament in general, and at nuclear non-proliferation in particular.

Two of the so-called threshold states, Israel and South Africa, exist in Africa and the Middle East. The perception of the threat to their national security (from within and without) has been claimed as an argument for the failure of these two countries to join the NPT. Other threshold countries inevitably follow the same policy. This demonstrates the incapability of the NPT to achieve universal denuclearization.

Egypt signed the NPT on 1st July 1968, and ratified it on 26 February 1981.

During the UN Special Session on Disarmament a resolution put forward by the Arab States was approved by 86 states and opposed by 3 (Israel, USA and Portugal) with 44 abstentions.

The resolution:

"Noting the persistent Israeli refusal to commit itself to non-proliferation in spite of the repeated calls by the UN General Assembly and IAEA, as well as the request of the Security Council calling upon Israel to place its nuclear facilities under the IAEA supervision;

Noting with deep alarm the recent information in connection with continuing nuclear activities, and considering this policy as a component of Israeli nuclear armament policy;

Noting the repeated condemnation of Israel for its refusal to abide to non-proliferation and of its continued cooperation with South Africa in this field.

Calls on all states and organizations that have not yet done so to discontinue cooperation with, and technical assistance to, Israel;

Requests the IAEA to suspend scientific cooperation with Israel".

Obviously, nuclear capabilities could not proliferate to regions of conflict without the collusion of nuclear states.

There is massive evidence of the roles of the USA, France and FRG in the acquisition of nuclear capabilities by Israel and South Africa. This has been confirmed in CIA reports as well as by the statement of an Israeli nuclear worker.

The inevitable deterrent and retaliatory response by threatened states, by chemical weapons and missiles, is not only due to the unbearable threat of mass destruction, but also to the extensive collusion by a superpower which allows Israeli military dominance over its neighbours. The measure of the Israeli-American collaboration is indicated by the launching of the Shavit booster rocket with a pay load of 2420 lb (capable of carrying a nuclear warhead) that can reach as far as Moscow. Furthermore, Israel also launched Jerico II with a range of 900 miles. These two major events, coupled with the revelation by Mordechai Vanunu about the Dimona nuclear plant, and the Israeli possession of 100–200 nuclear warheads, indicate the degree of insecurity among neighbouring states.

The uneven favouritism of the USA in supporting and propagating Israeli military nuclear monopoly and dominance in the region, is not only a hindrance to a peacefull settlement of Middle East problems, but it is also a booster for an arms race with conventional and strategic weapons.

In addition, the current status of the Israeli chemical industry assures it a full capability to produce all modern chemical weapons and their delivery systems.

Those regional dimensions of the NPT are rapidly becoming the determinant factor for future survival, unless the realities of the situation are faced.

Can the NPT Survive?

Nuclear proliferation cannot be dealt with in isolation from other weapons of mass destruction in the context of Third World regional security. Thus, it would necessitate the prohibition of any threat capability, whether by nuclear, chemical or conventional weapons.

If the NPT is to survive, new clauses will have to be added to stop the collusion and undermining efforts exercised by the superpowers, i. e. 1) assurance of stoppage of acquisition of nuclear weapons, their delivery systems, and other strategic weapons; 2) positive security guarantees to non-nuclear signatories; 3) access to all peaceful nuclear technologies and capabilities by non-nuclear signatories.

A nuclear-free zone, particularly in the Middle East, cannot be used as an escape hatch for those states who refuse to abide by international law and human rights in settling their disputes, and who use unlawful means to maintain a nuclear monopoly in the region.

Signing the NPT and accepting a peaceful settlement of the Middle East dispute are the minimum requirements for a nuclear-free zone in that region.

Nuclear-Weapon-Free Zones

Shalheveth Freier

Introduction

Any instrument which restrains the horizontal and vertical spread of nuclear weapons should be viewed as desirable. From the point of view of mankind as a whole, this statement is as trivial as the negation of war as a means of settling conflicts. The latter principle, enshrined in the Charter of the United Nations, is manifestly held in disregard. It remains an attainable but distant goal.

Restraint in the nuclear realm, however, commends itself to urgent and unflagging effort. In a war by conventional means, the parties to the conflict have time to assess the damage inflicted upon them as the war proceeds and call a halt when they deem it unacceptable. A nuclear exchange may inflict irreparable damage, allow no time for second thoughts, and break the taboo on the employment of nuclear weapons elsewhere.

Deterrence is often credited with having prevented war in Europe since World War II. This may be so, but deterrence cannot serve as a prescription for the rest of the world. Charitable hindsight can only conclude that by lucky circumstance there was a confrontation between only two blocs, and that the fear of the consequences of a nuclear exchange asserted itself even before deterrence became a concept and a policy. A spread of nuclear weapons would not meet the criterion of bi-polarity, and the incipient and uneven procurement of nuclear arms would lack the inhibition of deterrence, such as results from a comparable and parallel deployment on the part of two protagonists.

There are, therefore, good grounds for unceasing efforts to curb the spread of nuclear weapons. The workability of any international arrangement in this direction is of course encumbered by the fact that this is not a peaceful world, that the security of a country is not generally looked after by the international community, and that even such ambitions in the nuclear realm as do not seem warranted from a security point of view are not universally condemned.

It is natural, therefore, that one compromises in order to secure as large an adherence as possible, as is the case with the Non-Proliferation Treaty (NPT), or contents oneself with limited membership and makes allowance for individual exceptions, if conditions are more stringent, as is the case with nuclear-weapon-free zones (NWFZs).

However, summing up these generalities, the very challenge posed by institutional arrangements and the need for a country to declare in favour of a concrete commitment or otherwise, is by itself healthy. If a country declares in favour, the general trend towards restraint is encouraged. If a country holds back, the problems of the situation are brought into relief.

The Non-Proliferation Treaty

Before turning to NWFZs let us look at the NPT. Essentially, the non-nuclear-weapon-states (NNWS) party to the Treaty undertake:

- not to procure for themselves, or aid others in procuring, nuclear weapons, and
- to submit to IAEA safeguards all source or special fissionable material in all nuclear activities under their control.

The nuclear-weapon-states (NWS) party to the Treaty undertake:

- not to aid others in the procurement of a military nuclear capability, and
- to negotiate the cessation of the nuclear arms race and its reversal, with nuclear disarmament as a goal.

The membership of 139 states includes: NWSs; states which have nuclear weapons on their soil – though not under their direct control; states which are under the protection of NWSs; states which perceive no warlike threat to themselves; states which seek access to nuclear technology – irrespective of its ultimate purpose; and states which perceive no foreseeable nuclear threat in their neighbourhood. Despite this incongruous composition, the NPT has two achievements to its credit:

- Safeguards procedures have been put in place and are credibly applied by the IAEA, and
- a certain measure of international on-site verification has been accepted.

On the other hand adherence to the NPT:

- has not prevented a single local war;
- makes the acquisition of a military nuclear capability more difficult, but does not rule it out. (The IAEA acknowledges that the constraints on its inspection allow for uncertainty);
- has not led to a credible effort on the part of NWSs to reduce their nuclear armaments and thereby to narrow the gap between the NWSs and NNWSs, a division particularly odious to a populous country like India, on grounds of principle and circumstance;
- has not attracted the so called threshold states: Argentina, Brazil, India, Israel, Pakistan, South Africa.

However, if the present international climate reflects a sober awakening rather than a passing swell of hope, the two achievements of the NPT may turn out to have been a pioneering thrust. On the same assumptions, its under-achievements may eventually shrink in importance.

Nuclear-Weapon-Free Zones

The concept of a NWFZ is much more stringent than that of the NPT:

- It is a mutual commitment by states in an integral and inclusive zone to forswear the existence of nuclear weapons within the zone, and to withhold aid from states outside the zone in acquiring a military nuclear capability;
- It accepts the principle of mutual and reassuring inspection, in addition to such support as the IAEA may be invited to give;
- It requests nuclear powers outside the zone with dependencies or defence alliances within the zone to respect its nuclear-weapon-free-character;
- It arises from an initiative of the states in the zone and from free negotiations among them and receives at most international blessing on a job well done.

Precedent has shown the concept to be feasible. Its merits reside in the fact that:

- its provisions are meant to reassure neighbouring states directly concerned, and not a disparate international community, as is the case with the NPT;
- it affirms implicitly that even conventional wars are ruled out within the zone.

The limitations of the concept of a NWFZ arise, of course, from its demanding nature, as will be illustrated in the discussion of individual cases.

Treaties Establishing NWFZs in Populated Areas

1. Treaty for the Prohibitions of Nuclear Weapons in Latin America (Treaty of Tlatelolco)

The main obligations of the contracting parties include:

- no testing, use, manufacture, production or acquisition of nuclear weapons by the parties, or on behalf of anyone else;
- no receipt, storage, installation or deployment of such weapons, by the parties or on behalf of anyone else;
- no participation, elsewhere, in activity designed to produce nuclear weapons.

The Treaty, furthermore, maintains the right of the parties to conduct nuclear explosions for peaceful purposes, under appropriate control.

The provisions for verification and control include the establishment of a permanent zonal supervisory organ, OPANAL, to ensure compliance with IAEA safeguards and the additional provisions of the Treaty. In particular, provisions are made for "special inspections" in the case of suspect activity, upon the reasoned request of one of the contracting parties, and eventual engagement of the Security Council, in case of violations.

There is an additional Protocol I, which requests the states which have possessions within the zone (Netherlands, UK, USA and France, in this instance), to apply full-scope IAEA safeguards within their possessions, and to refrain from the manufacture, stationing and testing of explosive nuclear devices in them. There is an additional Protocol II, which requests the five NWSs to respect the character of the zone and not to use or threaten to use nuclear weapons against the contracting parties.

The Treaty is in force for 23 Latin American countries, some of which have waived their right to delay accession, until all states have ratified the Treaty and the two Protocols have been affirmed by the outside powers concerned. Brazil and Chile have ratified the Treaty but not waived the requirements, and the Treaty is not yet in force for them. Argentina has signed, but not ratified it yet. Cuba will not accede until the USA relinquishes its military bases on the island.

Protocol I has been ratified by the Netherlands, the UK and the USA, but not by France, which wishes first to ascertain the state of ratification of the Treaty. Protocol II has been ratified by all the NWSs.

It is noteworthy that the Treaty is in force for its 23 signatories despite the fact that Brazil, Argentina, Chile and Cuba have not yet assumed their obligations and that France has yet to ratify Protocol I.

Differences of interpretation and reservations exist with respect to:

- nuclear explosions for peaceful purposes;
- the permissibility or otherwise of the transit or transport of nuclear weapons;
- the applicability of the engagements of the NWSs under Protocol II, in the case of unfriendly acts.

Despite these uncertainties, as well as those pertaining to undetectable submarine movements by the NWSs, the Treaty may be said to have established a precedent on which other zones may be built. The reservations of Brazil and Argentina, generally classed as threshold countries, will be largely determined by the relations between them, their sense of priorities, the international climate, and their commercial interests. Except for the latter interests, the present situation holds promise.

2. South Pacific Nuclear-Free Zone Treaty (Treaty of Rarotonga)

Protest of French nuclear tests on the Mururoa Atoll in the mid-sixties, and proposals to dump nuclear wastes in the Pacific, fuelled a general revulsion against nuclear weapons, and, at the insistence of New Zealand and Australia, most members of the South Pacific Forum established the Treaty in 1986. Just as the zone of application of the Treaty of Tlatelolco extends into the high seas, the South Pacific zone extends over vast tracts of the Pacific and Atlantic Oceans, from the West coast of Australia to the western limit of the Latin American zone and is contiguous with the Antarctic zone in the South.

In addition to the provisions listed in the Treaty of Tlatelolco (see above), the Treaty of Rarotonga stipulates:

- no nuclear explosive device of any kind (not even for peaceful purposes);
- no provision of nuclear material or equipment to any state (NWS or NNWS) unless adequately safeguarded;
- not to assist or encourage the testing of any nuclear explosive device by any state;
- not to dump radioactive matter at sea and to prevent such dumping by others.

No special permanent secretariat is established to verify compliance with the Treaty. This function devolves upon the Director of the South Pacific Bureau for Economic Cooperation. Since only Australia conducts nuclear research of significance, and all states are essentially under IAEA safeguards, the situation differs from that of the Latin American zone. Complaints will be investigated on the spot, and referred to the South Pacific Forum as the last resort.

There are three Protocols pertaining to the NWSs:

Protocol 1 – similar to Protocol I of the Tlatelolco Treaty (outside powers with possessions in the zone to apply IAEA safeguards and refrain from stationing or testing weapons in them);
Protocol 2 – similar to Protocol II of the Tlatelolco Treaty (respect for the character of the zone);
Protocol 3 – no testing of any explosive nuclear device within the zone (i. e. on the high seas).

France, the UK, and the USA have declined to sign any of the Protocols. The USA, which signed both protocols of the Treaty of Tlatelolco, maintains that in the South Pacific instance a binding undertaking could limit its ability to meet its security commitments and undermine the policy of deterrence. France will not countenance a curtailment of its rights over its territories (e. g., the right to test). The UK's position is similar to that of the USA. All three, however, have made it known that they are in sympathy with the objectives of the protocols and will endeavour to act in accord with them. China and the USSR have signed Protocols 2 and 3, China unreservedly, the USSR with reservations.

The restrictions demanded of the NWSs on their freedom of movement and action over vast tracts of ocean have prevented the NWSs from affirmatively responding as they did in the case of Latin America.

3. Proposals for the Establishment of NWFZs in Populated Areas

A variety of proposals for NWFZs began to be tabled in the mid-fifties, of which we shall only say as much as is necessary for some useful conclusions to be drawn. In general, the USSR and the Warsaw Pact were supportive, while the USA and NATO stressed three conditions: the consent of all the parties to the zone (no UN imposition), no alteration of the military balance between the power blocs, and effective controls.

4. Denuclearization of Africa

In 1960, France conducted its first nuclear test explosion in the Sahara desert. Since then, the General Assembly of the UN has adopted yearly resolutions on the denuclearization of Africa: no procurement of nuclear weapons by African states or assistance to them in the manufacture of such weapons. There has been no action on this resolution by the African states, and since 1976, a yearly condemnation of South Africa, and repeated reports by the Secretary-General of the UN which testify to South Africa's technical capability to make nuclear weapons, have been the only concrete and concerted action by the African states.

To the outside observer it would seem that Africa could be an additional area – possibly the only one – in which the concept of a NWFZ could make headway. Apart from South Africa, there are only two states which could wish to retain nuclear options – Libya and Egypt. The major powers would not, presumably, stand in the way of a NWFZ for the continent. Coupled with the present climate of accommodation, in particular the new flexibility shown by South Africa, it is possible that the very negotiation on the modalities of a NWFZ for the continent could add confidence for the future evolution of relations.

5. South Asia

Following the Indian test explosion in 1974, India proposed in the General Assembly the creation of a NWFZ in South-Asia, arising from an initiative by, and negotiations between, the states of the zone. Pakistan, for its part, invited the states of the region and neighbouring NNWSs to initiate consultations to this end. Both resolutions were adopted by the General Assembly. Since 1976, Pakistan alone has continued to submit such resolutions, with India objecting on the grounds that no prior consultation had taken place among all the concerned parties on the feasibility of such a zone. India is concerned with its own great-power status in general, and with Pakistan and China in particular. The superpowers have preponderant interests in India and Pakistan, and remain aloof on this issue. It may therefore be doubted that a NWFZ is a feasible proposition in South Asia, especially as India sees no justification for it.

6. Central Europe, the Balkans, the Mediterranean, the Nordic States

A spate of initiatives, commissions and meetings since the mid-fifties in all these areas has not produced a feasible plan. These are the major areas of confrontation between NATO and the Warsaw Pact countries, and include members of both alliances. The concept of a NWFZ is not viable here, except in the context of overall arrangements aimed at arms reductions of all types and as an administrative corollary within such arrangements.

7. The Middle East

Ever since 1980 there has been a consensus in the General Assembly on the establishment of a NWFZ in the Middle East. The zone is not well-defined, but makes sense only if it includes all states which are wary of each other's intentions, i.e., from Libya to Iran. Unlike all other existing or proposed zones, this is one in which no state – except for Egypt – will countenance any negotiation with one member of the zone – Israel, the nuclear capability of which is generally assumed. The consensus resolutions depart, therefore, from the norm accepted for a NWFZ and seek to obviate any contact between Israel and the other states. They contain declaratory paragraphs and specifically insist that all nuclear facilities of all states in the zone be under full-scope safeguards. By virtue of IAEA safeguards on all installations, this would then be a NWFZ.

It is the Israeli view that free negotiation between the parties and *mutual* safeguards are the essence of a NWFZ, as is recognized elsewhere. The insistence on the application of IAEA safeguards to Israel, and the denial of negotiations with it, are interpreted in Israel as the Arab desire to bring Israel's nuclear activities under international control and retain the option of waging conventional wars against it without needing to worry about Israel's purported nuclear potential.

Israel, for its part, believes the modalities of the Treaties of Tlatelolco and Rarotonga contain those confidence-building elements – free negotiation and mutual reassurances – which will make the incidence of recurring wars less likely. It is these which worry Israel the most, and the increasing procurement of more far-ranging, more accurate, and more destructive implements of war by the Arabs, exacerbate these apprehensions. There has been no response to repeated Israeli invitations to the Arab states to negotiate a NWFZ and test the earnestness of Israel's intentions. At the behest of Egypt, both the UN and the IAEA have been requested to submit studies on ways to promote the establishment of a NWFZ in the Middle East.

In a region, packed with planes, tanks, artillery and rockets at a level four to ten times as high as that of NATO or WTO states of comparable populations – state by state – and as deeply rent by animosities as it is (Iraq – Iran, Iraq – Syria, Libya – Egypt, Libya – Sudan, Sudan – Egypt), even apart from the steady repudiation of Israel, a NWFZ strongly commends itself. It commends itself even more so as it is likely to have the approval of the major powers.

Conclusion

In the light of the foregoing, NWFZs are not really feasible in regions directly involved in the confrontation between NATO and the WTO, except as a corollary to more general arrangements on force reductions – men and arms, nuclear and conventional. This would be true of the initiatives for Central Europe, the Balkans, the Mediterranean and the Nordic States.

NWFZs are a feasible concept, in their own right, where they have been established (Latin-America and South Pacific), where all states of the zone do not really perceive a major threat – supported by a nuclear potential – on the part of their neighbours. There are two instances of hesitancy which may both yield to the improved international climate. These are Argentina and Brazil in Latin-America and the reservations of the superpowers about their freedom of movement and action. No NWFZ is conceivable in South-Asia.

Africa and the Middle East are worth on-going efforts. The insistence on negotiation between the parties on an issue so vital to all may be one means of trying to cajole the countries in a strife-torn region to at least prevent a further escalation of conflict, in the first instance.

We have to recognize the dependence of the viability of NWFZs on the climate which prevails between the superpowers. Also, no region is isolated from what goes on in its neighbourhood. We should not, therefore, exaggerate the achievement where it is on record. But we should appreciate, as we said initially, that any formal arrangement which is not utopian, but feasible, is by itself a good thing. It challenges the states to either commit themselves to restraint or else brings the problems of a situation into relief. NWFZs do that.

This paper does not deal with other relevant treaties, such as the Antarctic Treaty, which functions well, or the treaties pertaining to outer-space, the moon, the sea-bed and ocean-floor, which present problems different from the ones which apply to inhabited areas.

Nuclear-Weapon-Free Zones, Security and Non-Proliferation

*Julio C. Carasales**

In any analysis concerning the real value and future of the concept of nuclear-weapon-free zones, we should begin by considering the objectives a state wishes to attain when deciding to join such a zone. Usually two main purposes are mentioned: to increase the security of the state and to promote non-proliferation.

It seems to me that the second one, non-proliferation, is a secondary goal, and not the principal one. If a state has non-proliferation as its top priority, taking precedence over everything else, it could easily sign and ratify the Non-Proliferation Treaty. Nuclear-weapon-free zones, whatever the results regarding non-proliferation, are attractive to regional countries mainly because they think that their security is increased by joining the zone. The "UN Comprehensive Study on the Question of Nuclear-Weapon-Free Zones in all its Aspects" states this very clearly: "The premise upon which any nuclear-weapon-free zone must be based will be the conviction of States that their vital security interests would be enhanced and not jeopardized by participation."

In other words, if efforts are to be made to induce states to become members of a nuclear-weapon-free zone, those efforts should not be addressed to exalt the virtues and merits of non-proliferation, but should instead be directed to convince those states that their security would really be enhanced by joining a zone. The strengthening of non-proliferation would be just an additional benefit.

Let us examine then how a zone could increase the security of a particular state. This state could be either a small one with no significant nuclear capability, or a larger one with a fairly developed nuclear industry.

In the first case, there is no doubt that the security of the small state would be increased by belonging to a nuclear-weapon-free zone, particularly in relation to more powerful neighbours with a potential nuclear capability. The small state has, as a member of a zone, the assurance that its stronger neighbours will never acquire or develop nuclear weapons. Regarding weaker neighbours, the existence of the zone is really irrelevant. The zone would also be irrelevant vis-à-vis the nuclear weapon powers, because, if any conflict develops between the smaller state and the larger power, nuclear weapons would certainly never be involved.

The situation is not the same when we are dealing with a medium-sized state with a sufficient level of technical sophistication in nuclear matters that would enable it to consider, perhaps as a remote possibility but not an impossible illusion, the production of nuclear weapons. For such a country smaller,

* The author is expressing personal views, not necessarily shared by his Government.

underdeveloped neighbours do not pose a security problem, and, from this point of view, to belong to a zone is also irrelevant.

The dangers to such a state's security could come from two sources: a) neighbours in the region which also have a nuclear capability and b) nuclear-weapon-states.

In the first case, it seems obvious that the security of the state would be increased if, through the existence of a common zone, its neighbour or neighbours, as the case may be, were prevented from acquiring nuclear weapons.

But an alternative should not be ignored. To belong to a nuclear-weapon-free zone, a state has to sign and ratify a multilateral treaty, a treaty which includes provisions with which the state may not be in full agreement. The state could consider some of the prohibitions and controls imposed by the treaty as unnecessary, unacceptable, or too rigorous. If that state finds that its powerful neighbour or neighbours have similar reservations regarding the treaty, they could establish, among themselves, some kind of arrangement giving each other enough confidence and security without having to accept inhibitions and intervention by other parties or organizations. In other words, two neighbouring threshold countries could achieve mutual security without having to join a nuclear-free zone.

We should now deal with the relationship between a threshold country and nuclear-weapon countries from the point of view of the increase in security it would receive by joining a nuclear-weapon-free zone. This question is naturally closely related to the kind of commitments entered into by the nuclear-weapon states vis-à-vis a particular zone. What would the regional state gain in security by joining a zone? My answer given to this question is "very little, if anything".

Two treaties establishing nuclear-weapon-free zones exist in the world today, Tlatelolco and Rarotonga. The latter one, however, has had its protocols rejected by three of the five nuclear powers, and in general its provisions are weaker than those of Tlatelolco. Protocol II of the Latin American Treaty has been ratified by all nuclear-weapon states and is in full force.

What is the situation in practice? In Protocol II, the nuclear-weapon states agree to "fully respect" the Treaty "in all its express aims and provisions" and "not to use or threaten to use nuclear weapons against the Contracting Parties".

In reality, this high-sounding undertaking amounts to just an expression of good will. Why? Because:

1. Four of the five nuclear powers have attached to their ratification of Protocol II several so-called "interpretative" statements which are, in fact, real reservations, forbidden by Article 27 of the Treaty and Article 4 of Protocol II. Those reservations are designed to cover situations in which the respective nuclear power, by its sole decision, would consider itself not bound by its undertaking to refrain from the threat or use of nuclear weapons against a member of the zone. That means, therefore, that the regional state could never base the preservation of its security on mere unilateral declarations which their authors could evade at will.

2. The question concerning the threat of nuclear weapons is a very elusive and delicate one. If a non-nuclear-weapon state member of a zone is in open conflict with a nuclear power, is the existence of nuclear weapons in the area of hostilities, by itself, a threat to the security of the non-nuclear state? Naturally, the nuclear power would not openly issue a warning to its enemy about the future use of nuclear weapons, but is not the presence of nuclear weapons in its armed forces a sufficient menace? Has not the non-nuclear state concerned the right to feel its security threatened, when, due to a deficiency in the treaty establishing the nuclear-weapon-free zone, it is legally impossible to verify whether the nuclear power has violated the treaty, by introducing nuclear weapons in the region? On top of this, it is even impossible to obtain from the nuclear powers themselves an official public denial of the presence of nuclear weapons in a particular area because of their common policy: never to confirm or deny the existence of nuclear weapons at a specific time or place.

As far as I am concerned, if there is an area in which there is much room for improvement in relation to nuclear-weapon-free zones, it is the question of the commitments that the nuclear-weapon states are willing to undertake vis-à-vis the zones. Of course, each region presents particular problems and difficulties, and the attitude of the nuclear powers will not always be the same. But, in general, if they decide not to participate fully in the operation of a zone; if they are not willing to assume truly meaningful undertakings and not merely symbolic ones; if they want the non-nuclear states of the region to disregard their legitimate concerns and to curtail their peaceful nuclear programmes while they, the nuclear powers, refuse to take any steps leading to a more even balance of rights and obligations; if, in short, they decide to pay only lip service to the zones without undertaking any real commitments which would enhance the security of the countries of the region; if they remain aloof, selecting which zones they like and which they do not – and some nuclear powers do not seem to like any – the future of nuclear-weapon-free zones does not seem very bright.

The points addressed in this paper are not, of course, the only ones relevant to the future of the zones. What I wish to stress is that, in my view, the way to get new countries, particularly threshold countries, to join existing nuclear-weapon-free zones, or to establish new ones, lies not so much in trying to convince them of the value of non-proliferation as in showing them that their security would be enhanced if they became members of a zone and, to this end, a change of attitude on the part of the nuclear-weapon powers is essential.

A Nuclear-Weapon-Free Zone in Northern Europe

*Jan Prawitz**

Introduction

In the early days of the nuclear arms race, the concept of nuclear-weapon-free zones was introduced as a possible measure to prevent further dissemination of nuclear weapons to more states and to avert the risk of nuclear war in specified geographical areas. The earliest proposal, relating to Europe, was made in the United Nations in 1956. In 1957 the Rapacki plan proposed the permanent absence of nuclear weapons from all the territories in Central Europe.

In those early days two different but parallel approaches to military denuclearization were pursued: the global or open-ended, and the regional or zonal. The former started with the adoption by the UN General Assembly of the "Irish resolution" on the prevention of the further dissemination of nuclear weapons. That was the beginning of a process leading finally to the agreement in 1968 on the Treaty on the Non-Proliferation of Nuclear Weapons (NPT).

The NPT foresees in Article VII that "any group of states could conclude regional treaties in order to ensure the total absence of nuclear weapons in their territories". While two nuclear-weapon-free zones have been established in populated areas – in Latin America and in the South Pacific – most proposals for the creation of such zones have not materialized. This is also the case of the various proposals to establish such zones in different parts of Europe.

At the end of the 1980s, with the INF Treaty in force and a fifty per cent reduction of strategic nuclear weapons being negotiated, and the issue of a third zero-option in Europe being debated at political levels, the idea of establishing nuclear-weapon-free zones in Europe may soon be outdated or unattractive due to agreements on more far reaching measures. Such zones may, on the other hand, play a role and be tailored to fit a more general arms control pattern in Europe.

---

* Views and opinions expressed in this paper are those of the author and do not imply the expression of any position on the part of Sweden's Ministry of Defence.

History of the Nordic Zone Proposal

Among the various proposals for establishing nuclear-weapon-free zones in Europe, three still attract political attention: the Nordic zone, the Central European corridor and the Balkan zone.

The idea of establishing a nuclear-weapon-free zone in Northern Europe was suggested for the first time in January 1958 in letters from the Soviet Prime Minister to the Prime Ministers of Denmark, Iceland and Norway. The idea was followed up in several subsequent statements suggesting a "rocket and nuclear-weapon-free zone in the Scandinavian Peninsula and the Baltic area", as well as a combination of three proposed zones – Scandinavian-Baltic, Central European and Balkan-Adriatic – into a single nuclear-weapon-free zone from the Arctic to the Mediterranean.

The idea of establishing a nuclear-weapon-free zone in Northern Europe was, however, introduced by Finland. In May 1963, President Kekkonen of Finland suggested that the Nordic countries should establish themselves as a nuclear-weapon-free zone. Prospective member states were Denmark, Finland, Norway and Sweden. Mr. Kekkonen noted that despite the differences in their security arrangements – Denmark and Norway belong to NATO while Finland and Sweden are neutrals – none had acquired nuclear weapons and none had accepted those of other states on their territories. A Nordic zone would thus only confirm a *de facto* situation. Mr. Kekkonen's objective was to remove the Nordic countries from "the sphere of speculation caused by the development of nuclear strategy". It should be noted, however, that one of the prospective member states, Sweden, at the time entertained a nuclear weapon option policy that was not formally written off until 1968 when Sweden signed the NPT.

President Kekkonen returned to his zone proposal in 1978, now basing himself partly on experience from the implementation of the Latin American zone and on the results of an expert study undertaken by the United Nations in 1975 on the initiative of Finland. President Kekkonen's proposals were discussed from time to time during the sixties and seventies. The different attitudes of the Nordic governments prohibited, however, concrete negotiations on the issue.

A turning point was reached in 1980 for the Nordic zone proposal. A wave of new political interest emerged, beginning with Norway. A year later the governments of Denmark, Norway and Sweden expressed, with somewhat different formulations, a positive interest in the idea. New government-sponsored studies were initiated.

With a delay of a few years, this interest also spread to the "West Nordic area", i. e., Iceland and the Danish provinces of the Faroe Islands and Greenland. In 1984 the local governments of the latter two declared their territories nuclear-weapon-free zones. But as their foreign and defence policy remains the competence of the Danish government in Copenhagen, these "zones" have no legal status. They do show, however, that there exists an interest in the nuclear-weapon-free zone idea also in that area. One consequence of this development is that there are now two geographical zone-concepts under discussion. One is the original Kekkonen area, the other is the "large zone" with the West Nordic territories added.

Another conclusion is that a Nordic nuclear-weapon-free zone should not be established in isolation but in "its wider European context". First, Denmark, Iceland and Norway are members of NATO and would have to co-ordinate their policies with other members of the alliance. Secondly, because the Nordic area adjoins areas where nuclear weapons are deployed in large numbers, there is an obvious need for accommodation to its immediate military environment.

The attitudes of the nuclear weapon powers to the Nordic nuclear-weapon-free zone idea have always been mixed. The Soviet Union, which advanced the idea in 1958, has supported the proposal by regular official statements ever since. The Western nuclear weapon powers have never supported it and are an obstacle to the process, both generally and within NATO.

After years of discussion within the Nordic countries, the issue became ripe for consideration in 1987, when a joint Nordic governmental committee was established to "investigate the possibilities for the establishment of a nuclear-weapon-free zone in the Nordic area as a contribution to the efforts to achieve détente and disarmament in Europe." This joint investigation will last several years.

Geographical Considerations

The geographical concept of a Nordic nuclear-weapon-free zone has its basis in the tradition of cooperation among the historically and culturally close Nordic countries. The original Kekkonen concept comprised Denmark, Finland, Norway and Sweden – "the continental four" – and their territorial waters and air space. The large zone concept will also include "the Atlantic islands", i. e., in the first instance Iceland, the Faroe Islands and Greenland, but probably also the Norwegian island possessions of Jan Mayen and Spitsbergen with Bear Island.

The idea of including sea areas beyond territorial waters in a Nordic nuclear-weapon-free zone has not been discussed very much, with one important exception. The Swedish government, later followed by the others, has proposed that the Baltic Sea should be included in the zone. That would require a separate legal instrument because of the different legal status of sea areas as opposed to land areas. The Nordic countries have no jurisdiction over sea areas beyond territorial waters that permit them to declare such areas denuclearized. To be effective, denuclearization of sea areas must be based on agreement among the nuclear-weapon states rather than the zonal states.

It should be noted that the Nordic area includes three small areas that are demilitarized according to old treaties. They are a narrow zone around the southern part of the border between Norway and Sweden (1905), the Norwegian Spitsbergen Archipelago in the Arctic (1920) and the Finnish Aaland Islands in the

Baltic (1921). These treaties were concluded long before the atomic bomb was invented, but the areas could today be considered denuclearized as well.

While it is clear that the Nordic nuclear-weapon-free zone proposals should be considered in their European context, it is also clear that such zone proposals also have relevance for the discussion of naval arms control in general, and for the discussion of various regulation possibilities for the Arctic area.

Basic Elements

The achievement of the basic objective of a nuclear-weapon-free zone would require, in the general case, the adoption of three measures:

- non-possession of nuclear weapons by zonal states;
- non-deployment of nuclear weapons in the zonal area;
- non-use or threat of use of nuclear weapons against targets in the zonal area.

The meaning of those measures seems clear enough. However, their legal realization could be complicated as shown, for example, by the definition of a nuclear weapon in the Tlatelolco Treaty (Article 5).

The non-possession measure is already in force in the Nordic area, as all five Nordic states are parties to the NPT.

The non-deployment measure would primarily apply to the zonal states as far as land areas are concerned. Zonal states could not, however, by agreement among themselves, prohibit innocent passage (or transit passage) by vessels of nuclear-weapon states in their territorial waters. As mentioned above, denuclearization of sea areas beyond territorial waters would require a special legal regime for the area.

Related to the non-deployment measure is the "transit" of nuclear weapons through zonal territory. This issue became controversial when the Latin American zone was negotiated. The problem was solved by not being solved. Transit was left to the discretion of individual zonal states to grant or not to grant. The South Pacific zone has a similar transit provision. In a European case the transit issue is bound to be even more sensitive because of the abundance of nuclear weapons in adjacent areas.

The non-deployment requirement is also fulfilled in the Nordic area in peacetime due to unilateral policies. Finland and Sweden maintain a declared policy of neutrality, which by definition excludes the stationing of nuclear weapons in their territories. The NATO members, Denmark, Iceland and Norway, long ago made unilateral declarations that they will not accept nuclear weapons on their territories "in peace-time". A zone agreement would thus require the codification of existing policy, extended to cover all circumstances.

The non-use measure would be a commitment by states controlling nuclear weapons, in the present situation the five recognized nuclear-weapon powers.

Legally this provision has been given the form of separate protocols to the zone agreements. Reservations could not be avoided, just as in the Latin American case.

Consideration of the non-use measure should be made against the background of the ongoing negotiations on general negative security assurances at the Conference on Disarmament in Geneva. All five nuclear weapon states have made unilateral declarations that they would not attack, or threaten to attack with nuclear weapons, states that have no such weapons of their own or host others on their territories. These declarations are not co-ordinated and include some conditions and reservations. The USSR also generally pledged not to be the first to use nuclear weapons in a conflict. The reservations are linked to the question of whether a state could be member of a nuclear-weapon-free zone and a military alliance with a nuclear-weapon state simultaneously. This is certainly possible provided, however, that the two sets of commitments are not contradictory.

In the discussion about the possible establishment of a nuclear-weapon-free zone in Northern Europe, the idea was suggested in 1975 that the non-use element should be complemented by a "thinning-out" arrangement in areas adjacent to the proposed zone. Such nuclear weapons should be withdrawn as were targeted against the zone, or had short ranges and were deployed very close to the zone, thus making them usable primarily against the zone. If such weapons were not withdrawn, the non-use committment would be less credible. The Soviet Union has declared itself ready to discuss measures for thinning-out on its territory close to the prospective zone.

Concluding Remarks

In the sixties and seventies a critical question was whether a Nordic nuclear-weapon-free zone was really necessary, because it could upset a delicate balance between the two military blocks in Europe. Today the pertinent question is whether a Nordic nuclear-weapon-free zone is really necessary in the light on the ongoing deliberations and negotiations on more wide-ranging measures of nuclear arms control relevant to Europe. The proper answer to today's question is that the ongoing work on the issue is justified because it will provide political readiness should a Nordic zone be considered a desirable building block for the future European arms control system.

In a wider arms control context, various actor states would have different interests and different roles. The five small Nordic countries may consider a Nordic nuclear-weapon-free zone a regional security-enhancing arrangement in their favour and their contribution to a larger arms control pattern for Europe. The nuclear-weapon powers, on the other hand, might regard such a zone as a collateral measure with a possible confidence- and security-building effect on their general nuclear arms control system, a measure whose administration could be left to the Nordic states.

Part III

Nuclear Proliferation and Soviet-American Relations

The Treaty on the Non-Proliferation of Nuclear Weapons and the Relations Between the Superpowers

Ben Sanders

Throughout recent history, concern about the spread of nuclear weapons to additional countries has been a major consideration in the policies of nuclear-weapon states. As the Romans said: "Quod licet Jovi non licet bovi: what Jupiter may do, it does not behoove his ox to do". That is an interesting attitude, usually expressed in parental warning like: "Don't do [whatever you are doing], it is bad for you." It conveys a taboo, typically one-sided: "What you are doing is bad. I do it and can't stop doing it. But I shall do what I can to stop you." And as is the case with many parental warnings: it is entirely justified.

That is the paradox of nuclear non-proliferation, embodied in the Treaty on the Non-Proliferation of Nuclear Weapons, among other places. It runs like a red thread through the policies of all nuclear powers.

It all started with the original nuclear-weapon country: the United States. In the mid-Forties the USA was the only nation with the capacity to manufacture nuclear weapons. It had achieved that capacity with help from others, notably Great Britain and Canada. Almost as soon as the efforts to build the bomb became successful, American authorities took measures to ensure that no one else would achieve the same capability. They limited the involvement of non-Americans and, the close alliance of World War II notwithstanding, largely excluded Great Britain from nuclear policy making.

That policy was not only aimed at allies but at others as well. I do not commit myself here to the theory that the American government set out to establish a monopoly of nuclear military might. Many well-informed people take this view [1], but in retrospect it is hard to assume that the US government imagined it could make a secret of an elemental physical phenomenon, would be able to bar access to a relatively simple technology, and would retain a monopoly on the necessary industrial capacity. I am also unconvinced that the policy was primarily aimed at the Soviet Union. In those early years, relations had not yet deteriorated to a point that would have justified a non-proliferation policy aimed mainly at the USSR, especially given the generally low opinion America had at the time of Soviet science. Soon, however, the Soviet Union became the primary object of attention of America's nuclear policy, and this paper sketches the development of relations between the superpowers with respect to nuclear [non-]proliferation, from open hostility through grudging interaction to active cooperation.

It has been an instructive development. Throughout times of friction and mutual suspicion, the awareness of common interest in this area has grown on both sides. For more than twenty years – since well before the end of the first cold war period, and throughout the more recent tensions – the Big Two have cooperated: informally and unsystematically at first, and in the past few years, deliberately and methodically. This cooperation has become a factor in the increasingly amicable relations between the superpowers and may be relevant as a precedent for joint action in other areas of world concern.

The implementation of the policies arising from the common interest in nuclear non-proliferation shows some gaps. These may be expected, given the differences in the respective relations of each major power with third states, but they detract from the viability of a joint policy. There are instances where one or the other has given priority to a policy which appeared to override the non-proliferation argument. During the conflict in Afghanistan the USA has continued to give Pakistan military aid, although it was clearly embarked on a nuclear-weapon programme which under US legislation should have excluded such assistance. The USSR is assisting India in its nuclear-submarine ambitions, although that nation is a notorious hold-out from the non-proliferation system. In its supplies to the Indian nuclear-power programme the Soviet Union does not require the imposition of safeguards on all Indian nuclear activities, as most nuclear supplier states would do, but merely stipulates that the items supplied are so covered. The USA, again, seems to have turned a blind eye to nuclear developments in Israel.

Still, these are exceptions to the rule that there is a strong similarity of interest. Both states clearly recognize this. And although at times their respective styles of diplomacy prompt them to give different expression to their non-proliferation attitude (especially in their dealings with the developing world where the Soviet Union seems less intransigent than the USA) it would be a mistake to take this as a substantive difference. The USSR has made this obvious at several review conferences of the NPT, where it has at times cautiously supported criticism by non-aligned nations of the Treaty's implementation, but taken pains to ensure that those would not jeopardize the outcome of the conference.

A sketch of the relationship in nuclear non-proliferation between the superpowers over the last four decades must have as its background at least a rough outline of the history of that subject. Many of those involved in the Pugwash Movement are well-acquainted with that story, but newcomers would do well to take good note of past events. Trite or no, as Santayana said, those who cannot remember the past are condemned to repeat it. But strangely, although the history is so recent and so interesting, it is difficult to find objective sources. Some of the many books written on the subject seem to present a retrospective wisdom that is not always entirely credible. Several memoirs speaking of early events now strike one as self-seeking revisionism [2].

But good or bad, most publications pay insufficient attention to events in and around the United Nations. In hindsight, some of those events may not have been decisive. But they should not be ignored. They reflect the attitudes of the secondary players, smaller and newly independent nations that expressed themselves on the subject. Also, UN sources have the virtue that they are precise.

UN verbatim records reproduce exactly what was said, and if they do not always reflect the actual considerations of the country whose policy they pretend to convey, as public presentations of that policy they play an important part on the international stage [3].

The first atomic weapons exploded in August 1945, six weeks after the signing of the UN Charter. This put the UN before grave and unforeseen problems, such as how a body created to maintain international security, devised at a time when conventional military conditions prevailed, was to cope with the existence of a weapon many times more destructive than anything so far known; how it should take into account that only one state had that weapon (while others *might* eventually get it); and how it might adjust its decision-making procedures to the new situation.

In its first resolution, in January 1946, the General Assembly of the United Nations – arising from an initiative of the five permanent members of the Security Council and Canada – unanimously decided to establish an Atomic Energy Commission which was to make specific proposals for the elimination of atomic weapons from national armaments. At the first meeting of the Commission, the US delegate, Bernard Baruch, submitted the proposal to create an international atomic energy authority for the world-wide oversight of nuclear energy in all its manifestations. The Authority would not only have a regulatory function but would exercise "managerial control or ownership" of all applications of nuclear energy "potentially dangerous to world security" and have the power to "control, inspect and licence" all other nuclear activities. Shortly thereafter, the USSR submitted a draft convention outlawing nuclear weapons, but not regulating the peaceful uses of nuclear energy.

In late 1946, the General Assembly unanimously adopted a resolution which recommended, *inter alia*, consideration by the Security Council of a convention creating an international system of control and inspection, prohibiting atomic weapons and ensuring the use of atomic energy exclusively for peaceful purposes. Shortly after, the Security Council adopted (with the USSR and Poland abstaining) the first report of the Atomic Energy Commission which endorsed the idea of a treaty creating a strong system of international control and inspection, operated by an international authority with extensive powers, to ensure that atomic energy was used only for peaceful purposes. A few months later, the Soviet Union proposed amending these recommendations: it would have the international inspection, supervision and management apply to all existing plants and place over-all authority for managing the system in the Security Council.

From there on, the positions hardened. The USA adhered to the idea of an international body that would own all nuclear source material and the installations where fissionable material was produced and processed; be responsible for the control and oversight of the development of atomic energy; and have a great deal of autonomy. Its decisions would not be subject to the veto of the Security Council. The Soviet Union was opposed to the intrusiveness of such controls and to the idea of international ownership of materials and facilities. It wished the international body to be in the framework of the United Nations, subject to decision-making by the Security Council.

In the next few years, the contradictions deepened. The USA continued to push its plans for international control. The USSR advocated the primacy of a convention outlawing nuclear weapons, without which, it held, the establishment of international control would make no sense. The fronts continued to stiffen until by late 1948 the Commission had reached an impasse. Discussions of that time in the General Assembly reflected irreconcilable differences. The majority supported the US approach, while a group of states around the Soviet Union saw it as ensuring the continuation of the American nuclear monopoly and called for the prohibition of atomic weapons as a corollary of any international control. Consultations between the six Permanent members of the Security Council, convened to solve the impasse in the Commission – which had not met since the preceding summer – were unsuccessful. The fourth session of the Assembly, in 1949, started with the news that the Soviet Union had exploded an atomic weapon. In early 1950, the USSR withdrew from the six-power talks, which were not resumed.

The early years of the next decade were characterized by a multitude of proposals, raised in several bodies set up in the framework of the United Nations in the hope that new fora might promote new understandings. Most of the proposals dealt comprehensively with both atomic and conventional disarmament, although around the mid-fifties both sides and their allies, as well as third states, submitted a number of proposals for partial steps. None of these were yet addressed specifically to the question of nuclear proliferation. And none brought the major powers closer to agreement.

But there was one event that would have major consequences for future non-proliferation moves. With the end of the discussions on the Baruch Plan, the idea of an international authority that would control all nuclear materials was no longer on the table. That had in fact been an attempt at non-proliferation through ownership and control. In further attempts, the disarmament aspect would be lacking, but the non-proliferation idea would live on, especially in Western, and specifically American, thinking. It was given new impetus in proposals made by President Eisenhower to the United Nations General Assembly in December 1953. These proposals, the "Atoms for Peace Plan", were based on the idea that the peaceful uses of atomic energy could be furthered by the transfer of nuclear material from military to civilian uses. An international agency would be created through which all cooperation in nuclear matters would be channelled. That agency would have at its disposal stocks of nuclear materials to be allocated to peaceful pursuits, thus absorbing material hitherto earmarked for military purposes, which would be a form of arms reduction. The plan did not mention the international verification measures to be applied, but the first sketches for the statute of the international agency foreseen in it already contained that organization's right to apply "safeguards" and verify the material allocated to any state. Thus, the non-proliferation element came back into the picture. At the time Eisenhower spoke, however, it had become obvious that a number of states, beside those that had detonated explosive devices (by then the United Kingdom had also done so), had gone in for atomic development. It was to be expected that even if the International Atomic Energy Agency would become the main supplier of nuclear materials and promoter of atomic development, and in so doing would extend its safeguards to a large part of the world's peaceful nuclear activities, some activities might already have escaped the safeguards net [4].

The Statute of the International Atomic Energy Agency, which had required almost three years of preparation, was opened for signature on 20 October 1956. Its safeguards provisions had formed a very sensitive part of the negotiations. The Soviet Union had stayed away from the early preparations but had a lengthy exchange of notes with the United States, in which once more it insisted that no international agreement on the control of atomic energy could be viable without a prohibition of nuclear weapons. It also made it clear that it wished the atomic agency to be closely connected with the United Nations and be subservient to the Security Council, and that it should not become the exclusive channel through which states might cooperate on peaceful nuclear projects [5].

In the second stage of the preparations, representatives of six governments, all scientists, convened (during the closing days of the first Geneva Conference – the "International Conference on the Peaceful Uses of Atomic Energy" – in August 1955) to discuss the issue of safeguarding peaceful uses. Again, no noteworthy progress was made. The occasion was nevertheless important, as it was the first time since the mid-forties that the United States and the Soviet Union engaged in substantive discussions on the issue of nuclear non-proliferation. And from then on, the USSR participated actively in the preparations for the IAEA.

My own, indirect, involvement began with the activities of the Statute Conference, which discussed a draft worked out by representatives of twelve governments. As a junior member of the nuclear team at the Ministry of Foreign Affairs in the Hague, I read the reports of the Dutch delegate and had to make the first drafts of his instructions. One vivid impression was how little a delegate of a country like mine could contribute, and how frequently one read that a particular phrase was "a delicate compromise, resulting from long and arduous negotiations". One soon learned to take this as a warning: "Hands off – the big boys agree – don't rock the boat". For a young perfectionist, new to the game, this was often hard to take.

Another lively recollection is that the safeguards provisions in the draft Statute kept getting watered down. At the time, this was largely the result of actions by the USSR and its friends, who sought to reduce the intrusiveness of international verification. Efforts to reduce the stringency of the Agency's safeguards system and its implementation have continued ever since, but as time has gone by, the Soviet Union has changed its minimalist policy to a growing advocacy of strict and effective safeguards [6].

The development is characterized by a series of events which hindsight shows to have signified an important evolution in Soviet policy. Early pronouncements of the Soviet member of the Agency's Board of Governors – when that body, in 1959, first discussed safeguards in connection with a small supply of nuclear material – seemed to show an altogether negative attitude to the concept of international verification. The United States, on the other hand, strongly supported the Agency's safeguards role. The result was a series of particularly unpleasant debates, which to those in the Agency's Secretariat did not seem to bode well for the Agency's safeguards. Some time later, when the IAEA was requested to assist another state in the construction of a research reactor, calling for safeguards, a similarly acrimonious debate took place. For several years thereafter the issue would arouse acrimony between those two delegations, and their exchanges did not just reflect the bad relations of that time but singled out international verification as a particular bone of contention.

The differences appeared to have narrowed by 1963, when the Board discussed the first formal safeguards system, applying to research reactors. And a few years after that, when it devised and extended a system of wider scope (the "INFCIRC/66" system, still used in states not party to the NPT or the Tlatelolco Treaty), the USSR cooperated actively in its development. In 1965, the authorities in Moscow for the first time made a senior Soviet expert available for recruitment into the Secretariat unit charged with the development and deployment of the new system: a first-rate technician whose positive attitude towards the Agency's safeguards work demonstrated the change in Soviet policy. From then on, Soviet cooperation with the Secretariat in safeguards matters grew ever closer, and it became obvious that its policy in that regard had begun to run parallel with that of the United States, with which it now seemed increasingly to agree in the Board of Governors and in technical meetings. In the early 'seventies there still were obvious differences between the two countries' safeguards approaches, as reflected, for instance, in their different attitudes to IAEA activities in nuclear-weapon states. At about the time the NPT was opened for signature, the UK and the USA offered to submit their peaceful nuclear activities to safeguards, principally to encourage others to do so as well, by demonstrating that, although not obliged to do so, they were also willing to share the presumed inconvenience of international inspection [7]. This was criticized by several member states, including at first the USSR, as an unnecessary burden on the IAEA that would add to the cost rather than to the cause of non-proliferation. The USSR did invite technical staff of the IAEA to nuclear installations of the type it exported abroad, to help them work out verification measures, but it was made clear that such visits should not be seen as inspections. Then, in the early 1980s, the Soviet Union unexpectedly also accepted IAEA safeguards on selected peaceful facilities. Together with the USA and several other Western countries, it now also gives increasingly active support to the IAEA in developing and improving safeguards methods and techniques.

I have traced this evolution at some length, because its significance strikes me as going beyond its impact on IAEA safeguards and even non-proliferation. I believe that it implies an increasing acceptance of the concept that the mutual trust which is a primary pre-requisite for international security in the modern world cannot exist without mutual openness – a new concept where the Soviet Union is concerned. And it shows an awareness that Soviet and American interests in this regard, if they do not altogether converge, at least run parallel.

In retrospect, the initial negative attitude of the USSR to IAEA safeguards may have been typical of a more general hostility to the idea of international verification altogether, arising from the traditional Russian concern about foreign intrusiveness. It is a moot question, whether the Soviet attitude was partly inspired by the thought that nuclear safeguards might be used as a precedent for international measures to monitor compliance with eventual disarmament treaties. I have no concrete evidence to that effect, but it is my opinion that it must have entered into Soviet thinking at an early stage. It is hard to imagine that an idea entertained by a number of Western governments and expressed by some – including my own – at the very first General Conference of the IAEA, would not have occurred to the authorities in Moscow.

One early factor in the Soviet Union's policy change may have been China. In the early years, the USSR made its nuclear exports almost exclusively to countries with which it was ideologically or militarily allied [8]. Such exports were not made subject to safeguards. However, they were generally limited to reactors; irradiated nuclear material was returned to the USSR for reprocessing; and Soviet experts were employed at the facilities in question, who could monitor the use made of the supplied installations and material. Under the circumstances there was no apparent need for "safeguards" as such [9]. This changed, however, after the political break with the People's Republic of China, in the late 'fifties, when China almost immediately developed a nuclear explosive capacity, in part, no doubt, through the use of nuclear supplies obtained from the USSR, which had been meant for peaceful purposes only. Rumour had it, that this came as a great shock to Moscow, but whether it was prompted thereby to take a more positive attitude towards international verification of disarmament measures is not clear. It certainly began to take a more benign look at the Agency's safeguards shortly afterwards.

Thus, over a period of about fifteen years, the policies of the two major powers with respect to the IAEA and its safeguards became ever closer, until now their cooperation has become a matter of course. One notes that long before the recent improvement in bilateral relations there were regular consultations between them on non-proliferation and safeguards. Their informal and pragmatic cooperation in those areas shows that the superpowers have long been aware of a shared interest, which seems to weigh more heavily than do disagreements between them in many other matters.

Is it right to judge superpower relations in the area of *non-proliferation* so much on the basis of actions and attitudes concerned with *safeguards*? I think so. International assurance of compliance with non-proliferation undertakings is a *sine qua non* for non-proliferation. The case of China was a painful demonstration to the Soviet Union how nuclear assistance may be diverted to military uses if there is no such assurance. Soviet interest in keeping German fingers off the nuclear trigger was another strong argument in favour of nuclear non-proliferation, at a time when the Western allies were discussing the creation of a multilateral nuclear force (MLF). International verification (in which the USSR would be directly or indirectly involved) would be an effective way to ensure compliance with a German renunciation of nuclear ambitions. Thus, non-proliferation and safeguards became accepted as two sides of one coin [10].

Meanwhile, between 1952 and 1957, the disarmament negotiations had gone on in the framework of the United Nations. They ended amid sharp controversy about substance and procedure. The next few years were bleak. Attempts to revive the negotiations in new fora failed. Talks in a newly created negotiating body, the Ten Nation Disarmament Committee, soon ran into a series of intractable East-West controversies; they ended in the Summer of 1960 when a particularly virulent crisis prompted the East Europeans to leave the Committee. Shortly after, however, the two major powers began consultations which by mid-1961 resulted in agreement on a set of principles that were to be the basis of negotiations on general and complete disarmament. Subsequently, they also agreed on the composition of a new body where these negotiations should be held: the Eighteen-Nation Committee on Disarmament (ENDC). When the ENDC convened, in early 1962, the USSR and the USA both submitted to it drafts of treaties on general and complete disarmament, in three stages. These drafts differed in important aspects but both stipulated that from an early stage in their eventual implementation, the nuclear powers should refrain from transferring control over nuclear weapons or information on their production to non-nuclear powers [11].

This approach fitted in with the activities or Ireland in the General Assembly. A few years earlier, the USA and the USSR had both made proposals in the Sub-Committee of the Disarmament Commission, to curb the spread of nuclear weapons. Their approaches differed basically. The Soviet Union, worried about West Germany, took a nuclear-weapon-free zone approach, by suggesting the creation of a zone of limitation and inspection of armaments in Central Europe. The USA held out for a treaty that would stop the spread of nuclear weapons by banning their transfer and receipt, with exceptions (which must have added to Soviet concern) for purposes of self-defence in the case of armed attack. Neither proposal led to immediate further action.

In 1958, Ireland submitted a resolution on the subject in the General Assembly. It was not voted on, but it set off a continuing debate. The following year Ireland had the issue inserted in the Assembly's agenda and submitted a draft resolution calling for an international agreement by which nuclear-weapon states would refrain from handing over the control of nuclear weapons to states that did not have them, and the latter would undertake not to manufacture them. The USSR was against – again, no doubt, thinking of presumed nuclear ambitions in the Federal Republic – because it did not forbid the transfer of nuclear weapons into the territory of non-nuclear-weapon states. The USA voted in favour.

In 1960, again, Ireland submitted a draft resolution along the same lines as in the previous years but now calling on both nuclear and non-nuclear-weapon states from that moment on to refrain from acts that would lead to nuclear proliferation. This time the USSR supported the draft; in so doing it specifically referred to the danger that would arise from giving the Federal Republic of Germany nuclear weapons [12]. In 1961 Ireland once more submitted a draft resolution, calling on all states to conclude an international agreement under which the nuclear-weapon states would agree not to help others acquire nuclear weapons and non-nuclear countries would undertake not to manufacture or get control of them. This was unanimously adopted, the first of a series that would lead to the NPT.

In parallel, and as a follow-up, to these activities in the General Assembly, the ENDC considered the matter. Progress here – and in the Disarmament Commission, which discussed the subject in 1965 – was hampered at first by Soviet objections to NATO plans for a joint nuclear force, which it considered incompatible with the concept of nuclear non-proliferation. This issue would long play a part in the negotiations: the USSR persisting in its demand that the treaty should prevent the transfer of nuclear weapons directly or indirectly to any non-nuclear-weapon state or military alliance; the United States arguing that the NATO plans did not foresee a transfer of control over nuclear weapons and thus did not contribute to nuclear proliferation [13]. Where the Big Two agreed, however, was in their opposition to a link between a non-proliferation treaty and other measures of arms limitation, as demanded in a memorandum from the non-aligned members of the ENDC.

A major development in 1966 was the introduction into the US draft of language carefully circumscribing the concepts of "transfer" and "control" of nuclear weapons. The new language, in fact, put an end to possibilities of joint nuclear activities in the framework of NATO. And it opened the door to agreement with the USSR.

In the Summer of 1967, the two superpowers submitted to the ENDC separate but identical drafts of a non-proliferation treaty. Still lacking was a provision on verification; this followed early in 1968, in revised versions of the identical drafts, providing for the application of IAEA safeguards. In March of that year, responding in part to proposals by non-nuclear-weapon states, the two powers again submitted a joint revised draft treaty, and once more, in May 1968, as a result of discussions in the General Assembly, they jointly made some revisions in the text. Shortly thereafter the Assembly "commended" the Treaty, which was opened for signature on 1 July 1968.

From that point onward, the superpowers have consistently worked together to promote adherence to the NPT and strengthen prospects for its eventual survival. Their cooperation withstood an early test during the Conference of Non-Nuclear-Weapon States, in the Summer of 1968, when they were subject to sometimes severe criticism. It is not widely known that the Soviet intervention in Czechoslovakia, just prior to the Conference, which led to American protests elsewhere, did not seem to disturb the relations between them in Geneva. As the IAEA's safeguards man-on-the-spot, in touch with both, I was particularly conscious of this. Similarly, at the three NPT Review Conferences which followed the Treaty's entry into force at five-year intervals, there has been nothing in the statements by either to remind one of the initial disagreements or of present disputes on other issues. The only divergence in approach of which I am aware was at the second NPT Review Conference, in 1980, when in the informal attempts to work out compromise language on the implementation of Article VI, regarding disarmament, and specifically, a nuclear test ban, the USSR delegate seemed to favour a non-aligned formulation over that supported by the USA. But, as stated before, this may have been a matter of style and tactics as much as of substance.

Recent developments have shown the importance of non-proliferation for the relations between the superpowers. Obviously, for instance, the INF Treaty would be unthinkable without the NPT. It is in the interest of both – and of world security in general – that the spread of nuclear weapons should be deterred. Non-proliferation must not be held hostage to the perfectionism of those who would barter consensus on its benefits for uncertain progress in other disarmament areas. Such progress has to come in large measure from the Big Two, to whom non-proliferation is not just a desirable condition but a *conditio sine qua non*. Those who accept the non-proliferation system as a positive factor in international relations enhance their own security and contribute to the cause of disarmament.

References

1 For a recent statement of this kind, see Sir Mark Oliphant, "Three Men and the Bomb", *Bulletin Atomic Scientists*, (March 1989), pp. 41–42. He says literally: "Beneath the rhetoric [of the Baruch Plan] most non-Americans detected a *Pax Americana*, with the United States the sole possessor of nuclear weapons"

2 A recent book by McGeorge Bundy, *Danger and Survival-Choices about the Bomb in the First Fifty Years*, (New York: Random House, 1988), is a notable exception. I have used it extensively, both as a source for new knowledge and to check and correct things I thought I knew

3 I have made extensive use of a series of publications put out by that organization, including "The United Nations and Disarmament 1945–1970", United Nations Publication Sales No. 70.IX.1; "The United Nations and Disarmament 1970-1975", United Nations Publication Sales No. E.76.IX.1., and the "United Nations Disarmament Yearbooks", which have been issued since 1977

4 See the SIPRI monograph, *Safeguards Against Nuclear Proliferation*, (Cambridge, MA: MIT Press, 1975)

5 See Paul C. Szasz, *The Law and Practices of the International Atomic Energy*, published in 1970 by the IAEA in Vienna as no. 7 of its legal series, on preparatory work on the Agency's Statute

6 The call to minimize the intrusiveness (as well as the cost) of safeguards is now heard mainly from non-aligned nations

7 This was spoken of, at the time, as "the equality of misery". The move was primarily intended to show the more highly developed non-nuclear-weapon states that other industrial nations did not fear the risk of industrial espionage supposed to arise from safeguards

8 Possible exceptions are research reactors, of which several were supplied to developing countries, including Egypt which, at the time, had close relations with the USSR

9 Most of the verification activities of the IAEA arose from agreements under which it took responsibility for safeguards to ensure compliance with agreements between a supplier state (in most cases the USA) and a recipient. Since USSR agreements did not provide for bilateral safeguards, it did not become a party to any "safeguards transfer agreement". Now, when it makes nuclear supply exports, if the recipient is a party to the NPT, they are subject to safeguards pursuant to that Treaty, and when the customer is a non-party, such as India or Cuba, the recipient is required unilaterally to submit the supplied items to safeguards

10 An illustration from the Conference of Non-Nuclear-Weapon States, held in Geneva in 1968: On that occasion, the Soviet delegate, Professor V. Emelyanov, introduced me to the USSR member of the Eighteen Nation Disarmament Commission, Amb. Roshzhyn, with the remark (in English) to the effect that: "Mr. Sanders is a member of the safeguards staff in the IAEA Secretariat, who held out in favour of safeguards when we were still opposed. We should be grateful. If he and his colleagues had not continued to fight for safeguards against our wishes there would be no Non-Proliferation Treaty now!"

11 See USSR: Draft Treaty on General and Complete Disarmament Under Strict International Control, Article 16 ("Prevention of the further spread of nuclear weapons") and United States: Outline of Basic Provisions of a Treaty on General and Complete Disarmament in a Peaceful World, Stage I, Part C ("Nuclear Weapons"), paragraph 4 ("Non-Transfer of Nuclear Weapons"), reproduced in the *United Nations and Disarmament 1945–1970, op. cit.*, as Appendices II and Appendix III, respectively

12 The USA, while expressing its support for measures designed to deter proliferation, abstained, *inter alia* because the resolution did not recognize the central responsibility of the major powers to halt stockpiling nuclear weapons

13 Dealing with relations between the superpowers on the issue of non-proliferation, this paper largely ignores the actions and attitudes of non-nuclear-weapon states. There were, in fact, a number of non-nuclear-weapon states that played an important part in these events

The US–USSR Relationship and Its Likely Impact on Nuclear Proliferation

Valerii Davydov

The beginning of a process of superpower accommodation on regional conflicts creates large opportunities for mutual action against countries which are seeking to create nuclear weapons. The improvement in Soviet-American relations leads one to hope that a dangerous neglect of nuclear non-proliferation in crisis regions will be prevented.

In the 1980s the effective solution of the problem of nuclear proliferation was seriously hindered by Soviet-American political and military rivalry in regions where the main candidates to the "nuclear club" are situated. Despite the fact that annual bilateral consultations on nuclear non-proliferation began in the early eighties at the height of Soviet-American rivalry, and despite official recognition of the significance of this problem for international security and for Soviet-American interests, non-proliferation policy found itself in deep crisis. This is highlighted by the fact that all states candidate to the "nuclear club" showed considerable progress in accumulating and mastering nuclear technology and materials suitable for military use. Moreover, the military plans of these states clearly show the desire to acquire reliable delivery vehicles, starting with missile technology. Official explanations in the United States of the need for SDI, for international measures to control the export of missile technology, and to prevent nuclear terrorism, are accompanied by references to the growing threat of the spread of nuclear weapons. It is surprising that despite Soviet-American rivalry, no state has openly declared itself nuclear. But this should not be too surprising, because the majority of threshold states decided that it was better not to provoke the founders of the nuclear non-proliferation regime – the Soviet Union and the United States – by their actions, but to proceed quietly with their business, laying the technological base for the possible acquisition of nuclear weapons. The strategy of "bombs in the basement", initiated by Israel, proved to be beneficial both from the political and from the military point of view for the majority of threshold states. Moreover, this strategy was also acceptable to the instigators of Soviet-American rivalry, since it created no obstacles to their geopolitical game.

The distraction of the superpowers, rooted in their bipolar confrontation, led to their losing all effective levers of influence over the threshold states' policies by the end of the eighties. Even worse, these states learned how to profit from Soviet-American rivalry. Today it is even more difficult to determine clearly the stage of development of the nuclear programme of this or that country. The problem lies not only in the top secrecy of these programmes. This kind of "sensitive" question is considered to be irrelevant in relations with friendly states; it is simply put on the shelf or not mentioned. Calculations on how to use these states in some global combination against the other side dominated other considerations of the proliferation threat.

If this trend continues to develop, the leading nuclear powers will at best find themselves political hostages to "young" nuclear countries, and at worst may well be involved in a local nuclear conflict against their own will.

A frequently pursued pragmatic approach to the prevention of nuclear proliferation – the view that the development by a particular country of several nuclear bombs falls short of a full-scale nuclear force and is, therefore, not dangerous – could turn into a long-term miscalculation of unpredictable size. This short-sighted pragmatism can lay the ground for a strategy of allocating new nuclear states to one's respective alliance. That was the case with Europe in the fifties and sixties with NATO's nuclear members, Great Britain and France. Great Britain and France can be called reliable allies of the United States only to a certain extent. But this reliability can hardly be expected from some new nuclear country, because all of them follow, or try to follow, an independent course on the international scene. It is regrettable that signs of such pragmatism are evident in the approach to the majority of threshold states.

Is there a politically clear alternative to such a course? If the answer is "yes" then what should it be? It seems that if an effective nuclear non-proliferation policy is possible, it must be based on two axioms:

- First, nuclear proliferation is not a lesser but rather a greater threat to Soviet and American security interests than are their own opposing nuclear forces. This conclusion must find practical realization in the existing political relations of the two countries in any crisis region of the world where threshold states are situated. It seems that the political relations between the Soviet Union and the United States must be restructured from the point of view of preventing nuclear proliferation.
- Second, the United States and the Soviet Union must recognize that it is hardly possible to contain further nuclear proliferation if the threshold states are viewed only as pawns of their policy. A solution is possible only on condition of close cooperation between the superpowers and the threshold countries, taking into account their legitimate security interests. At present it is impossible to match the task of containing nuclear proliferation with the attempt to preserve indefinitely the privileged position of states which already have nuclear status in the political field.

It seems that we can hope for success in non-proliferation policy only when it is based on these principles. The present situation in the non-proliferation field certainly calls for an unprecedented degree of Soviet-American cooperation which should have begun long ago. Will the United States and the Soviet Union be able to restructure quickly their political relationship from the point of view of preventing nuclear proliferation? The success of non-proliferation policy depends to a great extent on the answer to this question. The whole fate of the non-proliferation regime, and the threshold states' challenge to it, depends on the solution to another problem – whether the United States and the Soviet Union will be able to realistically examine their own interests with due regard of the threshold states' interests.

The Middle East

The creation of a nuclear-weapon-free zone can play an important role in slowing down the proliferation of nuclear weapons here. All the states in this region, including Israel, support the idea. The General Assembly of the United Nations has regularly adopted resolutions supporting it. The USSR continues to underline the necessity to create it. The USA is not in principle against a nuclear-weapon-free zone in the Middle East. But in the absence of a general Middle Eastern settlement, the issue of a nuclear-weapon-free zone cannot pass into the sphere of practical action.

The majority of Arab states consider the nuclear-weapon-free zone to be a means to stop Israel's nuclear preparations and to put all Israeli facilities under international control. At present it is difficult to say whether Israel is ready to agree to this demand. But the ruling circles in Israel understand that the Arab states are not likely to allow Israel to have an indefinite nuclear monopoly. Israel is also very much troubled by the fact that some states in the Middle East are becoming interested in acquiring chemical weapons as a kind of counterweight to its nuclear potential. It seems that under these conditions Israel can be positive both on the question of a nuclear-weapon-free zone as well as that on a chemical weapon-free zone.

The fast dissemination of ballistic missiles could turn this region into a potential nuclear and chemical threat to the USA, Western Europe and the USSR. In order to stop these dangerous trends, far-reaching cooperation between the countries of the West and East is needed. The global acceptance of the obligations in the Soviet-American INF Treaty acquires particular significance. The USA and USSR, as well as West-European countries should undertake an initiative to ban all intermediate- and short-range missiles everywhere. With a view towards preventing nuclear proliferation, it is also very important to ban and dismantle all nuclear weapon production facilities.

Regarding a nuclear-weapon-free zone, the USA and USSR should guarantee its nuclear-weapon-free status – not to violate it, nor to use or threaten to use nuclear weapons against members of the zone. Moreover, the nuclear powers should commit themselves to the security of all nations in the zone. Such actions demand a critical reappraisal of American and Soviet Middle Eastern policy.

South Asia

India does not accept the idea of a nuclear-weapon-free zone because of the proximity of the nuclear weapons of China and the United States (in the Indian Ocean). It considers a guarantee not to use nuclear weapons against a non-nuclear country an unacceptable slight to its national dignity. But the question remains: could India agree to a nuclear-free-zone on condition of the partial denuclearization of the Indian Ocean and of Chinese territory bordering on India?

From the viewpoint of a general approach to the problem of retarding nuclear proliferation, the situation can be greatly improved by a further normalization of the international climate in the region. Nuclear powers (USA, USSR,China) should do their best to explore possibilities in the realm of nuclear-non-proliferation. This deserves the highest priority on the agenda of a possible summit meeting of these countries.

The Afghanistan settlement enlarges American options for coping in a positive way with Pakistan's nuclear ambitions. The "Symington amendment" could play a crucial role in this policy. The USA, USSR and other states, which supply modern arms to the region, could contribute towards the non-nuclear use of these weapons by imposing special provisions in bilateral arms sale agreements.

Finally, the USA and the USSR should try to play a more active role in mediating the process of further normalization of the Indian-Pakistani nuclear relationship.

Africa

The establishment of a nuclear-weapon-free zone in Africa has a very good chance of success. Almost all the states support the idea. Beginning in the sixties, the General Assembly of the United Nations regularly adopted resolutions in favour of it. At the 1964 summit conference, the Heads of States and Governments of the Organization of African Unity (OAU) issued the Declaration on the Denuclearization of Africa. The USSR supports a nuclear-weapon-free zone. The USA does not officially object to it. The nuclear-weapon-free zone could create a substantial political barrier against the nuclear ambitions of South Africa. Implementation of the idea of a nuclear-free zone will inevitably provoke the question of international control over South Africa's nuclear facilities. Pretoria's attitude to this question may be considered as an important indication of that country's nuclear intentions. It also permits a test of South Africa's recent declarations to join the NPT. Even if South Africa refuses to participate in an agreement on a nuclear-weapon-free zone, progress in this direction would positively influence its nuclear ambitions. It is important to begin practical action towards the realization of the international status of a nuclear-free Africa. Both the USA and USSR should cooperate to achieve this.

The US–USSR Relationship and Its Likely Impact on Nuclear Proliferation

Cesare Merlini

Introduction

They say that from the beginning of time small and large states have equipped themselves with the best arms they could afford to achieve or defend their interests. As long as international relations are based on substantial anarchy (and some believe it is best that way because the only alternative, they say, is an empire) it is inevitable (and perhaps opportune) that nations have or try to have the instruments with which to command respect. If that is the way things really stand, why waste time and energy in the quixotic (or imperialistic) battle for nuclear non-proliferation? Would it not be better to establish some kind of international order that includes the many large nuclear powers, but also the medium and maybe even the small ones? This question begs another: is it really true that a world with many militarily nuclear countries would be a more dangerous and unstable place than it is today? Nuclear weapons could provide a more effective defence, discouraging aggressors with their lethality; as someone figuratively put it, they would be the spines of a world of porcupines, peace-loving but unassailable animals [1].

If deterrence works for the superpowers why not reproduce it in various regional contexts? Small nuclear forces would allow countries to establish their own equilibria, reciprocally dissuading each other from attacking, and to use the bomb as a final resort against larger nuclear powers. This theory has already been applied to Israel in the proposal that the government in Jerusalem should openly declare its nuclear weapons and the conditions under which it would use them, in order to permanently dissuade the Arabs from any idea of annihilating the State of Israel and at the same time to provide a basis for the start of negotiations with them [2].

This is an example of "the dove's dilemma". If nuclear weapons can prevent conflicts, is it not better that they exist? This dilemma concerns both horizontal proliferation (the porcupine model) and vertical proliferation, in that mutual deterrence among nuclear powers generates peace.

Edward Teller is certainly not a dove. His position has always been that since the proliferation of nuclear weapons cannot be halted, it would be better to question the current conviction that nuclear war cannot be limited. As early as 1960 he wrote that the danger of a limited nuclear conflict triggering a total conflict was minimal, which is why we have "to continue the development of light, cheap and flexible tactical weapons" and spread knowledge about nuclear weapons [3].

In Europe, NATO followed Teller's advice and instead of undertaking a more expensive strengthening of conventional defence, acquired thousands of these light, cheap and flexible tactical bombs (around 7000 at the beginning of the seventies) as a convenient low-cost solution, until it realized that it was damaging the stability of the balance and the effectiveness of defence.

This property of nuclear weapons – lower cost for the same destructive power – also underlies another question. If they are so cheap, why deprive the poorer nations of the possibility of creating their own nuclear defence with which to dissuade their enemies as the rich nations do?

In a nutshell, the attempt to prevent the spread of nuclear weapons to new countries is 1) anti-historical because it goes against the natural evolution of weapons; 2) inopportune, because it does not allow nuclear deterrence to extend to the regional level; 3) unfair, because it prevents new countries, and in particular the poorer ones, from acquiring a nuclear defence and deterrence, which is, among other things, not very expensive.

Undoubtedly, non-proliferation policies can be defined as anti-historical or "unnatural" [4]. The question remains whether nuclear weapons are so novel that they exclude the application of some historical models. In other words, while the most powerful weapons available have always, from the beginning of time, been employed, only since August 1945 have weapons existed that can in one fell blow destroy vast territories and with repeated use perhaps jeopardize the survival of the human race, bringing the weight of such a risk to bear on the aggressor – even in the event of victory – thereby recommending other forms of action and exercise of power. The threat of use (rather than use) of nuclear weapons is one of these forms, although associated with other factors (delivery systems, communications and control systems). Therefore, first of all more powerful weapons were produced; then their yield was decreased, and they became more sophisticated in an attempt to graduate and specialize their effect. But that is as far as their "normalization" goes. Those who, like Teller, imagined that nuclear weapons would enter into the normal management of international conflicts, were mistaken, at least until now.

Despite "miniaturization", nuclear devices have kept their specificity. A threshold remains that separates them from conventional weapons, even if the latter now have greater destructive power. Contrary to Teller's logic (and, indeed, that of many Soviet strategists), this threshold has not been crossed, just like the people at the party in Bunuel's film *The Exterminating Angel* who do not venture out the door although there is apparently nothing to keep them from doing so. Despite disquisitions on the controllability and limitability of nuclear conflicts, no one has yet put them to the test.

The protracted existence of an unviolated threshold is of great importance in that the longer it continues, the more remote the possibility of repeating the

Hiroshima experience, the more uncertain the effects of multiple nuclear deflagration in a conflict and the more unpopular the idea of using nuclear weapons becomes. Their utility, therefore, resides not in their use, but in the threat of their use, that is, their deterrent effect. Nuclear weapon states (NWSs) possess this deterrent ability, but not all to the same degree. The United States' and the Soviet Union's are equal (more or less), independent and global. France's, Great Britain's and China's, since they have limited second-strike capabilities, depend on the risk of escalation drawing in the superpowers. This has to be kept in mind when reproducing deterrence in theatres other than the East-West is contemplated.

Any other new deterrent capabilities are bound to be not only local, but presumably unbalanced and increasingly dependent, in the sense that they are credible only to the extent that they can involve larger actors. Since deterrence among NWSs is not static, but extremely dynamic, due to the evolution of technology and its consequences on strategy, this situation is bound to become more rather than less marked. For example, China has obviously lost ground in the race with the superpowers.

If the outcome of the use of nuclear weapons by NWSs is highly uncertain and their deterrent effect is accompanied by limited independence, it follows that countries cannot hinge their national security exclusively on their nuclear capabilities. The latter therefore becomes an extra capability over and above their conventional one and in the end is not a cheap substitute at all, but a complementary luxury, as France and Great Britain demonstrate.

European countries in NATO found out first-hand that placing too much national defence in the hands of the so-called nuclear theatre weapons, deployed in answer to the numerical advantage of the Warsaw Pact's conventional forces, generated a situation of greater uncertainty. A costly reassessment of conventional forces was undertaken after 1978, coinciding with a reduction in tactical nuclear weapons, with deterrence entrusted to intermediate-range nuclear missiles. That was the picture until the latter were eliminated in the INF agreement after Reykjavik.

On the other hand, Israel has strengthened its security and has been able to pursue an expansionist policy thanks to extremely efficient conventional forces. Nuclear warheads have little to do with Israel's imposing military strength with respect to her neighbours. The fact that local deterrents are not independent of the larger NWSs', not only undermines the theory that nuclear weapons can endow a new member of the atomic community with greater autonomy, but also complicates the East-West balance and, therefore, negotiations on nuclear weapons. These negotiations are being carried out mainly between the two superpowers; the other NWSs – France, Great Britain and China – have not participated so far. Although their nuclear deterrent forces are not a decisive obstacle to an agreement because of their inferiority with respect to the United States and the USSR, they nevertheless complicate negotiations. Although it can not be proven, it may be deduced that the emergence of new nuclear powers that are either independent of or somehow linked to the superpowers, would make negotiations on nuclear arms control even more difficult.

Nixon recalls that during his presidency he contemplated the use of nuclear weapons three times [5]: 1) in the Indo-Pakistani war in 1971, which could have drawn in China (on Pakistan's side) and the USSR (on India's), 2) the Arab-Israeli war in 1973, when American forces were put on alert in response to a Soviet threat of intervention, 3) in moments of crisis on the Chinese-Soviet border. It should be added that Kissinger denied that these options were ever seriously contemplated at an operational level, above and beyond what "President Nixon thought in the privacy of his office or his home" and claimed that the only real case of actual risk of nuclear conflict remains the Cuba crisis in 1962 [6]. It is interesting to note, anyway, that in all three cases indirect conflicts or tensions between the two superpowers involved non-autonomous (China) or potential nuclear deterrents (Israel, India, Pakistan). In no case was the area of direct contact between the two alliances, Europe, involved.

The Dilemmas of the Protectors and those of the Protected

During negotiations on the NPT, there was a debate on the problem of the guarantees that NWSs would have to offer NNWSs in exchange for their renunciation of nuclear weapons. They included the former's commitment not to attack the latter with those weapons (negative guarantee) – a commitment which ended up as a vague declaration made in the UN Security Council, and later, as far as Latin America is concerned, in the signing of supplementary Protocols to the Treaty of Tlatelolco – and the promise to provide nuclear protection in case of aggression (positive guarantee) – a promise which forms the basis of several alliances, the main ones being NATO and the Warsaw Pact.

Although never formalized in a non-aggression pact, no NWS has ever used or threatened to use nuclear force against a NNWS. Both the United States in Vietnam and the Soviet Union in Afghanistan lost their respective bloody wars without being able to resort to even a minimal part of their nuclear arsenals. The precedent of the Tlatelolco Protocols, in particular the second, which obliges the NWSs not to use or threaten to use nuclear weapons against signatories, is important because it means that possible reproduction of the nuclear-free zone formula in other areas would not only prevent local proliferation, but could also provide a formal guarantee to countries adhering to it.

It is usually said that Western Europe is under the American nuclear umbrella and Eastern Europe under the Soviet one. Although already under an umbrella, Great Britain has its own nuclear force, and so does France. The latter has not integrated its *force de frappe* with NATO's, but benefits from the American commitment in the framework of the Atlantic Alliance. The other countries in the

Alliance, the Federal Republic of Germany, Italy, and others, have formally renounced nuclear weapons in the NPT, but have obtained a positive guarantee which has also been translated into deployment of American forces on their territory. Like hostages ensuring respect of the guarantee, they have both a political and a military function. Japan is in a similar situation. It is linked to the United States by a bilateral alliance, the only difference being that no American nuclear forces are stationed there (the very possibility of the stop-over of an American ship carrying nuclear weapons in a Japanese port has been reason for domestic controversy).

Germany, Japan, Italy, Belgium, and the Netherlands have long reached a technological level permitting them to equip themselves with nuclear weapons if they so desired [7]. No wonder they were recalcitrant when negotiations on an agreement against proliferation started. On the one hand, the Alliance to which they belonged constituted a substitute for the guarantee provided by their own deterrent. The NPT explicitly consolidates an inequality among its participants, codifying the renunciation of some states in return for a guarantee from others, and thus, in traditional terms, represents a loss of sovereignty. With their eventual signature of the Treaty, they shattered the front of NNWSs which had formed during preparatory negotiations. The situation in Eastern Europe was similar, even if the way in which it was achieved was different due to the lower degree of autonomy in foreign policy of the USSR's satellites.

Thus, alliances with the superpowers were an extremely important factor in non-proliferation. In fact, critics say that the NPT is no more than a reflection of the situation determined by the existence of the two alliances: all countries adhering to it are either directly or indirectly (such as the neutrals like Sweden and Switzerland) protected, or have no chance of ever achieving a nuclear capability for themselves or for others, while all the countries that have this potential – the threshold countries – and that do not belong to nuclear alliances, did not sign.

When substituting its own deterrent capability, guarantees of protection limit the country's independence, in that they transfer to the guarantor the decision on if, when and how to intervene. More substantial, however, than negative guarantees, which do not limit independence, they can protect against third parties in as much as the deterrent of the protector is effective.

Western Europe is familiar with the dilemma of being protected. On the other hand, it would like the United States to be willing to "sacrifice New York for Paris", because that is the only thing that can really dissuade the Soviet Union from attacking, and upsets every time any doubt is cast upon Washington's loyalty. This occurred, for example, when Kissinger, in a frequently quoted speech delivered in Brussels in 1979, warned the Europeans not to rely too heavily on the American commitment. His intent was to stimulate the United States' partners to take a greater share in the alliance, but his words were music to the ears of those who, like the orphans of Gaullism, claim that the Old Continent must defend itself more autonomously.

The commitment to protect calls for credibility and, therefore, the operational readiness of the US deterrent, in particular, of the nuclear weapons that European countries have requested or accepted on their territory. The basis for all this, naturally, has to be not only good will or respect for one's commitments, but a common security sphere: the American perception that its national security includes the defence of Europe (and Japan). This perception has held up till now against isolationist drives and can be explained by the fact that although extended deterrence is felt to be a costly burden, it does put the United States in an enviable position. Furthermore, there is an unwritten clause in the Alliance which states that in return for protection, European states must give up pursuit of the global strategic capabilities they once had (and which twice in history made military intervention by the United States necessary).

On the other hand, Europeans fear the Americans' inclination to "press the trigger", before having exhausted all political avenues for solution of conflicts. This could lead to a nuclear exchange without Europe's security having been undisputedly shattered. The so-called "dual key" system is an attempt to recover sovereignty: one key is in the hands of the Americans and the other in the country hosting the nuclear warheads. The latter obviously has no influence whatsoever on the positive decision to use the nuclear weapon, which remains the prerogative of the President of the United States, but it does have veto rights with respect to use. In reality, this dual key system poses some complex problems in the event of its actual application in case of conflict. Political problems, above all: although the scenarios for use of nuclear warheads in Europe are defined by NATO's Nuclear Planning Group, the final decision has to be preceded by a (multilateral? bilateral?) assessment of the situation which must be carried out in a very short time. And technical problems: the keys are made of magnetic card built into the warheads, and of complex and secret codes. Analysts agree that keeping up communications would be the greatest difficulty the Alliance would have to face upon the opening of hostilities. Moreover, the "dual key" system, if it exists, gives the host country the chance to prevent the use of NATO nuclear warheads but is impotent concerning the use of American warheads deployed in Europe [8]. Take the past debate on the deployment of Pershing 2s and Cruise missiles, for example. Some governments, including those in London and Rome, said they were holding a key, while others, like the one in Bonn, refused it.

Western Europe's limited sovereignty (freely chosen and accepted, unlike the limited sovereignty imposed on the East and theorized by Brezhnev, but limited sovereignty nevertheless) has been well compensated. The alliance's deterrent effect has ensured more than forty years of total peace with relatively limited defence expenditures. European economies have benefited from this and have grown remarkably, more than the United States'.

These guarantees of protection could be extended to new countries, thus extending the relatively cheap benefits of peace and security and thereby probably also extending prevention of nuclear proliferation. Even if it has caused the

United States other more serious concerns, the fact that Cuba – not a signatory of either the NPT or the Tlatelolco Treaty – is under the Soviet sphere of influence has reduced the risks of proliferation in that country. The influence and control exercised by Moscow over Libya and North Korea, now both signatories to the NPT, are well known.

These cases, which are in reality isolated, could be extrapolated to a situation in which the world is split into two alliances centered around the two superpowers, each protecting its allies (or satellites) from each other with their (positive) nuclear guarantee. This would be a non-proliferating world. It might even be a more peaceful world, as local conflicts would be subjected to greater control; in fact, if they were inside one of the two blocs, the hegemonic superpower would act as peacemaker; if they occurred between countries in the two blocs, the superpowers would try to prevent or at least contain them to avoid escalation into a general nuclear conflict.

This scenario of a world totally divided between two nuclear umbrellas is obviously not credible. It is true that the two alliances have not lost members up to now; on the contrary, they have gained them (except for the withdrawal of Albania from the Warsaw Pact long ago). Moreover, the group of non-aligned countries is less compact than ever, but it is also true that the capacity for control of local situations by the USA and USSR has decreased compared to the early post-war era. Controlling half the world each would not only call for strategic, political and diplomatic capabilities that are indeed increasingly out of reach of the two superpowers, but also for the abolition of the medium powers, whether aligned or not.

Providing a guarantee of nuclear protection implies a double cost for the superpowers: the economic cost of a deterrent that is strategically complete – in the sense that it satisfies all deployment requirements and has second-strike capabilities – and is geographically extensive – in that it covers all the allies' territory; and the political cost of exposing one's own population to risk in order to defend one's allies. Both superpowers have repeatedly complained about the excessive economic costs they have to bear, especially since the allies have become more affluent.

In Moscow, the feeling that it can no longer support the burden of military commitments of a guarantor has brought about an agonizing reappraisal which we will come back to later.

Polemics about burden sharing are recurrent in the Atlantic Alliance. When they become acute, like now, the United States exerts pressure by insisting that American troops stationed in Europe will have to be brought home in order to cut costs. It is evident that American public opinion is also caught in the dilemma between hegemony and isolationism. Positions between these two extremes do not necessarily reflect traditional political positions of right or left, conservative or progressive, Republican or Democrat. They vary from imperialistic hegemony to the quest for a polycentric Western leadership and for common values, and from super-armed isolationism à la *High Frontier* [9] to a semi-pacifist isolationism. (Besides these costs for the guarantors, there are also the costs borne by those guaranteed: the economic cost of participating in the common defence – which is less than that of the guarantor – and the political cost of a lack of total autonomy in security decisions and, therefore, in decisions in general – which is greater than the guarantor's.)

American hegemony, more oriented towards leadership than imperialism in the past, has generated a sense of protection among Western Europeans at an acceptable cost. Under Reagan, the imperialistic characteristics of American hegemony were accentuated. Thus, the political cost of the guarantees also increased. But there was no substantial change in the guarantees themselves, despite the uproar about a space-based defence. European doubts about security were actually alerted more when Reagan suddenly took the guise of the negotiator.

If a more isolationist position, be it leftist or rightist, were to prevail in the United States, European security perceptions would likely change. The same would presumably happen in Japan. *"Plus ca change, plus c'est la même chose."*

The situation in which the superpowers find themselves – guarantors within their respective blocs amid a majority of non-aligned states, many of which are poor or developing – is, despite appearances, relatively simple with respect to the situation preceding the great upheaval of the two world wars. First, the balance of power system, which dominated in the nineteenth century, was more complex, both in terms of the number of actors on stage and of the quality of their military forces and, therefore, of their spheres of influence (with associated guarantees of protection). The global and deterrent nature of today's nuclear weapons is greater than that of the military capabilities up to the first half of the century.

With a forty-year record, mutual nuclear deterrence can boast a certain stability. That does not mean that one should be fooled about its technical and strategic characteristics which are subject to change.

The story of how the brief American nuclear monopoly after the war was replaced by the threat of Mutual Assured Destruction is well-known. Offensive devices ended up prevailing in the superpowers' dreaded arsenals as a gauge of vulnerability and, therefore, deterrence.

After the qualitative leaps made in nuclear warhead technology with the invention of the powerful H-bombs (in the fifties) and very sophisticated warheads (in the sixties) the technical evolution that produced or accompanied this strategic evolution mainly involved delivery (aircraft, submarines, multiple-warhead ballistic missiles, non-ballistic missiles), early warning (satellites) and command, control and communications systems (computers). The result has been the growing accuracy with which small targets can be hit and the ever-decreasing time needed for decision-making and response.

The increase in accuracy has had an important consequence: targeting of nuclear weapons has shifted from (extensive) civilian targets (like cities) to (smaller) military targets. The underlying reason is not humanitarian; it is strategic: destroying the enemy's second-strike capability. Effectiveness and practicability are enhanced and so is the temptation to strike first.

The greater vulnerability of nuclear weapons initially stimulated research on ways to protect them by hitting incoming missiles with defensive missiles (anti-ballistic missiles). But given the high cost and limited efficacy of this kind of defence, the two superpowers came to an agreement and later signed the ABM Treaty.

This treaty could not and did not prohibit further research on defensive techniques. After a decade, and with the benefit of simultaneous progress made in the fields of satellites, computers and directed-energy weapons (high kinetic energy pellets, particle beams and lasers), this research has produced its fruits and certainly generated new expectations, especially in relations to the Strategic Defence Initiative.

This multi-layer defence system would have consisted of:

1) a network of hundreds of surveillance satellites to keep enemy territory and part of the oceans under observation;
2) a system for processing the data so obtained and for response control;
3) a second satellite network of killer satellites to hit enemy missiles during launching or in flight.

The idea that a system of defence against ballistic missiles would make nuclear weapons "impotent and obsolete", as Reagan stated in 1983, turned out to be a propagandistic expedient to create consensus for an extremely costly venture. It was, as such, rather successful. Initial consensus may have been facilitated by a large number of books, articles, films and scientific analyses on the terrifying consequences of nuclear war (such as "The Day After" and "Nuclear Winter"), on the possibility of the near extinction of the human race, on the uncontrollability of a conflict, on the risk of a war by error, on the moral need to find an alternative. This campaign, which originated prevalently among American liberals, helped to create an ideal atmosphere for the proposal of an anti-nuclear defence, which originated on the conservative side.

Not only was the defence potential blown out of proportion in order to create consensus, but the whole initiative was immediately brought out into the open with fanfare and trumpets, before the results could be ascertained. Normally, weapons are developed in secrecy until there is a reasonable certainty that they will work. With SDI it was exactly the opposite.

As a result, nuclear weapons are not obsolete. The composition of arsenals may change as a result, shifting from an offensive-defensive mix towards more of the latter with a different breakdown among delivery vehicles and – it is to be hoped – more decision-making time during a crisis. It is not clear whether that will reduce the probability of war. In fact, the widely held conviction that a balance, or a set of balances, is more stable if the opposing forces are more defensive (to the extent that forces can be defined as defensive) is not substantiated by experience.

A second result has been the development of a new generation of weapons focused on the target with reduced collateral damage. Satellites seem to be most endangered by these developments in military technology, which may be highly destabilizing in view of their role in intelligence and surveillance [10].

These rather ambiguous hypotheses have raised important questions for the medium powers, in particular, West European ones. What would happen to their security and the Alliance's strategies? What credibility was left for the two national nuclear forces in Europe?

The Americans tried to be reassuring. The new prospects, they said, strengthened the guarantee to Europe because:

a) they increased the second-strike capability of the United States and, therefore, its deterrent credibility; and
b) they called for the development of technologies that the Europeans could imitate in order to build complementary defences against Soviet weapons aimed at the Old Continent. But, the Europeans objected, by reducing the superpowers' vulnerability (it was probable that the USSR would also have undertaken some kind of defence), the deterrent balance that has backed peace in the last forty years could be upset. There would be the risk of nuclear war in Europe, which could be separated from the global context (the famous decoupling!), especially if short-range or "battlefield" nuclear weapons were to remain on European soil. Finally, there would be a higher risk of a limited conventional war.

Contrasting opinions have always been expressed about the usefulness of second-rank nuclear deterrents. They were restated during the SDI debate. On the one hand, "decoupling" would corroborate those who have always claimed that European countries cannot leave their defence up to the guarantees of a foreign state forever; the English and French deterrent forces would require costly up-dating and could possibly be integrated with their own mini-SDIs, but they would be autonomous. Had this position been confirmed, it would have found new supporters in those countries that gave up the nuclear option even though they have the technology: the Federal Republic of Germany, above all, and Japan.

But it could also have been claimed that "star wars" was about to mark the demise of the minor NWSs. The cost of up-dating the French and British deterrents (and even more so the Chinese) would have been prohibitive. Even if united into a scarcely credible very efficient Superstate, Europe would have difficulty in bridging the gap. The right course, then, was not to take part in the race at all.

Not only SDI, the hawks' favourite invention, but also positions on the opposite side of the American political spectrum created problems for Europe. At about the same time, they came out with ideas like "no-first-use", that is, a Western commitment not to use nuclear weapons in case of a conventional invasion by the Warsaw Pact (overturning NATO strategies) and "minimum deterrence", that is, reduction of the deterrent to a level that is more symbolic than strategic.

These contrasting stances all seem to express a sense of American "tiredness" with the strain of extended deterrence, a strain which is accentuated by the technological trend reducing decision-making times during crises.

The result is an increased risk of war by error, due to a mistaken assessment (human or technical) of enemy movements, or by a spiraling sequence of hostile steps and retaliations. The former is favoured by the growing rapidity and accuracy of nuclear weapon systems which are highly dependent on electronic intelligence; the latter by the presence of tactical nuclear weapons ready for use in the initial phases of a conflict. Both are facilitated by the existence of nuclear-equipped third parties (this last consideration explains why one of the points of convergence of the various American positions is non-proliferation).

If that is the real danger, "an agreement to cut current nuclear arsenals in half could have beneficial effects, but it would not in itself significantly change the basic shape of the problem we face", as some analysts who, in the midst of hawks and doves call themselves "owls", observe [11]. They are in favour of arms control, of course, but what really counts is: 1) crisis control – rapid and clear communications between governments during crisis; 2) avoiding the use of nuclear threats in order to apply political pressure; and 3) availability of instruments able to explain the enemy's behaviour. Plus a clear distinction between nuclear weapons and conventional weapons and, above all, no increase in the number of nuclear actors.

The proposal to set up two "Centres for Reducing Nuclear Risk", one in Washington and one in Moscow, equipped with a mixed staff from both countries and directly linked to the crisis control rooms of the two capitals [12], led to the agreement signed in September 1987. It established an electronic system for binding notification of the launching of ballistic missiles and the transmission of other information "out of good will", but neglected to implement the most innovative aspect of the proposal (a mixed staff).

From a political point of view, the agreement to eliminate INFs was of great importance. Despite the limited numbers and restricted categories involved, it marked a long-awaited positive turn in nuclear arms control negotiations. Equally important is that the superpowers have finally accepted effective verification, opening their plants to inspection by the adversary. If this were to lead to acceptance of the principle of international verification in the future, the repercussions would be of great moment above and beyond bilateral relations between Washington and Moscow.

The question comes up again whether the future role of the French and British nuclear forces is going to change. A looser link between the European theatre and the global strategic picture will accentuate their dilemma of whether to integrate more fully with the American deterrent (British model), with the prospect of a further loss of profile, or strengthen their autonomy (French model) with the prospect of isolation (as in the Chinese case). And yet, since they have intermediate-range missile capabilities, i. e. the capability to hit the Soviet Union, which NATO is being deprived of, they may be seen as the natural replacement for the INFs, and their validity would thus be confirmed, although at a very low level of credibility for the reasons stated above. For a while it seemed that the new situation was about to enhance a higher degree of convergence – still the main "variable" in the hands of European decision makers – between France and Great Britain and with other European countries, above all Germany. The latter's perception of being the nuclear battlefield *par excellence*, a perception heightened by the remaining and "decoupled" artillery and short-range missiles, caused a state of *Angst* among public opinion. The last two types of nuclear weapons are often put into the same basket, called Short Range Nuclear Forces, but this is misleading. While German concerns are justified as far as artillery is concerned, the same cannot be said about missiles. The lengthening of their range would have the double advantage of strengthening deterrence through "coupling" and allowing for dramatic reductions in the number of warheads, as will be seen in the next chapter.

Current Developments and the Future of the Non-Proliferation Regime

Four main conclusions can be drawn from the lengthy introductory considerations above. Two are conceptual:

- the specificity of nuclear weapons remains. As a consequence, concepts related to nuclear weapons such as deterrence and non-proliferation cannot simply be transferred to other types of weapons like conventional, chemical or biological ones;
- the near-bipolar and global nature of the current nuclear deterrence also remains. Reproducing nuclear deterrence on other scales and in other areas has proved unfeasible so far. Thus, deterrence and non-proliferation are closely linked, possibly mutually conditional.

And two are historical:

- the "protectors", i.e. the superpowers – whatever the important differences between them – both seem to be finding it hard to ensure continuing protection to the same extent as before, mostly for reasons related to relative economic decline. And both are keen to maintain a nuclear non-proliferation regime based to a large extent on their protecting role;
- the protected countries do not seem, for the time being, to be very interested in the idea of getting rid of their nuclear protectors. Over the years those in the West have gradually come to support the non-proliferation regime more than most of them did in the early days; those allied with the Soviet Union were not allowed to develop an autonomous view on nuclear issues and have at present other problems to deal with in their relationship with the protector.

In the debate about non-proliferation, those countries which are customarily described as "problem countries" fall into nonprotected areas (with the possible exception of Israel and, perhaps, Cuba, that may be considered protected states, as will be discussed later). But there are "potential problem countries", which because of their technological level (and possibly covert activities) can go nuclear, credibly nuclear, in a shorter time span than most "current problem countries". All of them fall into the protected areas, even if they are neutral, like Sweden [7]. They are not "current problem countries" because (a) they eventually formally committed themselves not to go militarily nuclear; (b) their perceptions of security have not changed substantially ever since; (c) they have become either supportive of, or resigned to the non-proliferation regime.

Moreover, it may be interesting to note that by far the largest share of IAEA safeguards (and practically all IAEA/Euratom safeguards) are carried out in the (protected) "potential problem countries" because of the intense civilian nuclear activities that are taking place in most of them.

A number of important developments are taking place against this background, which can be summarily listed under the following major headlines.

1. The USA and the USSR have reached an agreement to reduce to zero the INFs located in Europe and seem to be half-way through to another agreement that would sizeably reduce strategic nuclear forces.
2. A broad and complex dialogue has been initiated between East and West in Europe that may, through negotiations or unilateral steps, bring about reductions in the conventional forces of the two alliances and probably also in the remaining nuclear forces, as well as improvements in economic and human relations.
3. The Soviet and other Communist regimes are attempting reforms along highly differentiated patterns and to extremely varying degrees. Those reforms are mostly domestic and economic in nature, but foreign policies are clearly affected.
4. Change is also evident in the Pacific rim, because of the emerging economic powers, particularly Japan, which is on the verge of becoming the world's number one economic power.

The consequences of these developments on the future of the nuclear non-proliferation regime in those regions where countries rely on nuclear protection from the superpowers will now be analysed.

Europe

Besides the traditional specificity of being the most important area of "contact" between the two blocs and, thus, the area with the highest concentration of weapons of any kind, continental Europe is experiencing the new specificity of being the area where three out of the above four developments have their major impact, while not being indifferent to the fourth. Implications in all fields – strategic, economic, political, cultural – promise to be far-reaching. Here we will confine ourselves to the strategic ones, indeed to the problem of nuclear weapons alone.

Subsequent to the zero-INF agreement, the debate in the West on the future of the theatre nuclear forces witnessed the emergence of two schools of thought. For one, further reductions must be worked out with emphasis on the denuclearization of buffer zones to be enlarged so as to lead to a drastic reversal of the present situation in Central Europe, from the area with the highest concentration of nuclear weapons to a nuclear-free zone. For the other school of thought enough has been done in the nuclear field (wrongly, some believe) and now priority has to be given to conventional force reductions; nuclear forces, on the contrary, require modernization, partly to respond to the inevitable technological developments, and partly to compensate for the consequences of foregoing intermediate-range missiles in the flexible response strategy of NATO.

The Atlantic Alliance seems to be steering towards this second line, with the addition of further cuts (whether unilateral or negotiated is not clear at the present stage) in nuclear artillery, already reduced by some 2000 warheads or more in installments since 1979. This addition – it has to be underlined – is not a minor one. Battlefield nuclear weapons represent an important chunk of nuclear forces deployed in Europe by both alliances. Their deterrent effect is limited insofar as they represent either the recipe for a limited nuclear war – which would not please the Europeans, i. e. the would-be protected – or the trigger for very early escalation to a strategic nuclear exchange – which would not please the would-be protectors, who in the end decide whether to push the button or not.

Let us look, for a moment, at the scenario implied by the other line of thinking. A denuclearized Europe does not automatically mean the disappearance of the superpower guarantee. In theory, the umbrella may be maintained at only the upper level of direct exchange between the American- and Soviet-based or sea-launched strategic forces. Such a de facto guarantee would be weakened dramatically, even if not switched from a positive to a negative one (as described before).

More importantly, Europe would once again likely become an open field for conventional conflicts, most probably generated by resurgent nationalist sentiments. The current evolution in the Communist countries, besides promising developments, shows signs which are inevitably of great concern. The reference here is to nationalist movements and ethnic turmoil, which may sow the seeds of instability in the changing pan-European landscape.

Thus, the solution of further reducing certain kinds of nuclear weapons stationed in Europe, parallel to the cutting of conventional forces, while maintaining credible nuclear deterrence is the right one and should be pursued boldly. Credible nuclear deterrence implies that the remaining nuclear forces – say something between 1000 and 2000 warheads out of the currently estimated 4660 on the NATO side – should not be frozen in the current configuration, but modified according to technological improvements and strategic requirements. After the zeroing of the INF, the new mix would see a zeroing or quasi-zeroing of nuclear artillery (depending on the outcome of CFE), some remaining free-fall bombs, a few hundred (not-so) – short-range missiles and a number of stand-off missiles to be introduced anew [13].

Two out of the three other official NWSs that make nuclear deterrence *nearly* bipolar are European. The existence and the ambiguities of the British and French nuclear forces have already been discussed. London and Paris have succeeded in keeping them out of the arms control picture so far, and Moscow has shown a remarkable degree of flexibility in this respect. But, as an Italian political figure said, "they are not on the moon" [14]. Once NATO has better defined the Alliance's strategy in the new context of decreasing numbers, Britain and especially France should become more active participants in nuclear arms control negotiations, and their national nuclear forces should enter the picture of reductions. At the same time, they should increase their complementarity with the NATO deterrent and possibly become more "European" in their vocation. This may mean that while authority on use would remain national, planning and targeting activities would be shared.

The remote prospect of a common nuclear deterrent linked to a future strongly united Western Europe is certainly not brought any closer by current developments in Eastern Europe. Besides the clearly declining perception of threat, the Community is already finding it hard to implement all the provisions of the Single European Act and may have difficulty in maintaining a political profile were neutral countries like Austria, let alone current Warsaw Pact countries like Hungary, ever to seek entry.

How relevant is such debate on nuclear forces by nearly all flags to the future of the non-proliferation regime? To the same degree that the existence of the non-proliferation regime is relevant to the debate – not very. This apparent paradox resides in the fact that the regime was set up gradually and, to a certain extent, successfully to control the spread of nuclear weapons, while nuclear arsenals on and around the Old Continent were being built up at a dramatic pace.

Does the regime require the European status quo? Or can it sustain the opposite trend of decreasing numbers in nuclear weaponry? Would not the regime be best served by the above-mentioned solution of a Central European nuclear-free zone? The rather profound change in security perceptions of the European countries that this prospect is likely to generate, and the resurgent risk of conventional conflicts of nationalistic origin, so familiar to European history, would probably bring to the forefront the "potential problem countries" (as defined above). Initially they would simply experience domestic and possibly foreign pressure to change their attitudes and policies towards non-proliferation and to take some intermediate steps.

Instead, the maintenance of theatre nuclear deterrence at a lower numerical level of weapons, if obtained through successive steps, each characterized by greater stability, and if accompanied by measures to reduce the risk of conflict by accident or misperception (e. g. by developing and/or multilateralising "crisis centres" of the type set up in Washington and Moscow), should reduce the "nuclear fever" in Europe.

Such a conceivable scenario – by no means an assured one – of gradual change in the European nuclear landscape, would combine the survival of the nearly bipolar nuclear deterrence with the scaling down of nuclear arsenals. The nuclear threat to neighbouring areas remains very unlikely (the famous "out-of-area problems", repeatedly discussed within and between Western Europe and North America, have never implied nuclear solutions). The non-proliferation regime would be strengthened. The 1990 NPT Review Conference will have to recognize that non-negligible reductions – negotiated or unilateral – have begun in Europe during the previous five years, consistent with Article VI of the NPT.

As is well known, non-proliferation is not high on the agenda of European capitals. Except for a few opinion groups prevalently sponsored by American circles, the subject is not of frequent and profound interest even among political scientists. The twelve members of the European Community, however, have set up a working sub-group on non-proliferation in the framework of the European Political Cooperation. The signing of the NPT by Spain has helped to reduce differences among the member governments.

The Twelve may thus develop a common position for the coming Review Conference. A highly positive contribution to the Conference would be an announcement by the French government that it is considering signing the Treaty before the 1995 deadline.

East-Asia and the Pacific

At an IISS annual conference in 1986, Henry Kissinger discussed differences between the European strategic outlook and the East-Asian and Pacific one, and suggested "that the security and foreign policy of the Asian nations, different though they are, are more similar to what European foreign policy used to be in the nineteenth century, more dependent on a perception of the balance of power, a greater emphasis than has been the case in Western Europe on the political and geopolitical element, less insistence on American reassurance" [15].

If this is true, an international regime like the one introduced by the NPT and other nuclear non-proliferation provisions is even more "unnatural" there than it is in Europe. In all other world regions where the "balance of power" scheme is dominant – South Asia, South-West Asia and Latin America – "problem countries" dot the scene. Taiwan and South Korea used to be listed among the "problem countries". Now they should be considered more as "potential" ones and as such, be put in the same category as Japan.

As has been said before, Japan's economic rise is the main international "variable" in the region. The political and strategic implications of such a rise are still unclear to the Japanese people themselves. The issue of globalism and/or regionalism in the orientation to be given their foreign policy is very prominent in the political debate. The direct and short-term implications of such debate for nuclear non-proliferation appear very marginal. Tokyo not only remains committed to the NPT, but has substituted past policies of foot-dragging with a more cooperative attitude, subsequently developing a line of conduct of its own. The industrial sector, which used to be critical of the technological constraints deriving from US non-proliferation policies, has now come to support an international code of conduct that generates a more stable framework for a potentially expanding international role.

As for the changes taking place in the Communist countries, Japan has so far been the most "cold-footed" of Western countries about Perestroika. After having made sure that the zero-INF agreement was global, i.e. it also covers the Soviet intermediate-range missiles deployed in Asia, it has followed the ongoing East-West dialogue, particularly some West European exchanges, with detachment. Japanese industry has been discouraged by the government from rushing to Moscow with even only a fraction of the entrepreneurial activism it has displayed in other areas. The "Northern Territories" issue, which has been put forward as a priority matter, seems to have provided an appropriate façade for a cautious if not diffident attitude.

Japanese immediate concerns about "relaxation of tension" are today less directed to the Soviet Union and to China than to the United States, with which trade frictions periodically arise and, more or less indirectly, affect security, as was the case in the FSX fighter aircraft dispute. A widespread concern in Japan is how to use its economic power to limit American strategic hegemony, which includes holding the nuclear umbrella over the land of the rising sun.

In the longer run, however, things may evolve differently. Japanese economic strength is already such as to make the ceiling of one per cent of its GNP (constitutionally put on defence expenditure) approach the current French, German and British defence budgets, without taking into consideration those R& D activities which are labelled peaceful but have direct or potential fallout in the military sector. Among the latter are the development of nuclear fuel reprocessing and plutonium purification, as well as research related to uranium enrichment.

Japan has a very advanced nuclear industry, and the fact that the nuclear power programme is being implemented at a slower pace than planned, may turn into an advantage for two reasons: first, because it allows subsequent power stations to absorb new safety arrangements so as to make them more reliable than others on the market; second, because the Japanese nuclear industry (possibly in close alliance with the survivors of the American one) may find itself in a strong competitive position internationally if new environmental concerns ("green house effect", etc.) in the 1990s make nuclear power an attractive option.

Moreover, economic strength and generational turnover are introducing a more assertive attitude into the Japanese leadership, which at one and the same time shows new capabilities to deal with the complex international issues of interdependence and signs of new self-confidence, nationalistic pride and resentment towards foreign partners.

Finally, temptations for Japan may not necessarily be in the direction of reconsidering the nuclear weapon option. As for other advanced countries, intermediate steps are conceivable, like building nuclear-powered submarines. The insular situation and the geostrategic position of the country would certainly not deprive the proponents of such a programme of substantive arguments. And, if countries traditionally devoted to non-proliferation, like Canada, do consider this option, the example would probably not be disregarded out of hand in Tokyo.

What, in the light of these tremendous economic, technological and cultural potentials, can exert such pressure on the Japanese as to upset their present cost/benefit analysis of the nuclear option (or simply their self-control)? One or more of the following scenarios: 1) a collapse of the non-proliferation regime, either regionally or globally or both; 2) increased strain with the United States, with a consequent (or simply concomitant) loss of confidence in American nuclear protection; 3) a new rise in the perceived threat from the two Communist giants in Asia.

The current debate in Japan about its international role must be seen against the background of these scenarios. It appears to me that a somewhat stronger regional orientation may be appropriate in order to put Japanese strength to use in producing more structured development and stability in the area. Strategic stability will be helped by the fact that neighbouring countries are – out of past but not forgotten historical experience – likely to be watching each other to prevent the rise of new hegemonic roles, and particularly new nuclear states. Japan knows that a change in its attitude towards military nuclear capability, following, say, the French model, would inevitably entail a similar change with the Koreans and possibly others. Its position is likely to remain similar to the German's for quite some time. This should help the stabilization of the non-proliferation regime, thus preventing the first scenario.

But Japan will most probably also keep growing as a global actor. Western cooperative instruments will require continuing adjustment to a scaled-down American hegemony and will have to be improved (for instance by strengthening the Euro-Japanese segment) without putting the global strategic balance into question. That would take care of the second scenario.

Finally, the current trend of improvement in East-West relations should be continued and made irreversible by binding agreements and institutions. Japan should display a more open attitude and be called upon to take a more active role in such fora as the United Nations and others. This should help prevent the third scenario.

The alternate solution of developing a nuclear-free zone located between the area covered by the Rarotonga Treaty and the area corresponding to ASEAN [16] is not quite convincing. The situation is to some extent different from the European one, but negative guarantees only would not prevent Japan, Korea, Taiwan and possibly other countries from feeling cut off from the security system within which they have grown economically strong. The "balance of power" would probably deteriorate into major conflicts, as happened in Europe in the nineteenth century.

Cuba and Israel

These two countries represent hybrid cases. The Middle East is not a protected region; on the contrary it is a typical "problem area", located on the borders of the East-West strategic equilibrium. Israel, always at the top of the list of "problem countries", does, however, enjoy a guarantee from the United States, possibly even stronger than the one granted the European allies. Cuba, also a non-signatory of the NPT, is just across the US border, but a member of the Warsaw Pact. Both countries, though in quite different situations and for quite different reasons, have problems with their protectors and are encouraged to show moderation.

Israel is considered a *de facto* NWS and as such is a powerful incentive towards proliferation. Recent discussions have centred more on its delivery capability than on its nuclear weapon capability, but the latter inevitably acts as a multiplier of the concerns related to the former. Were the Middle East to split into two spheres of influence (as the Camp David process seemed to imply for a while), the situation would have eventually revealed several similarities with Europe, with Israel playing the French or British role. Were Israel to fully develop its launching capability so as to reach Soviet territory, the similarities would then match the 1962 Cuba pre-crisis situation, duly reversed.

Ever since the 1960s, Moscow has exerted control over Havana's nuclear ambitions, as already noted. Washington has not been able to exert similar control over Tel Aviv, whose nuclear activities (only military) have become a permanent handicap for the very active and occasionally dogmatic non-proliferation policy of the protector.

No major change in this complex situation can be expected. Pressures on Israel are numerous today, but have purposes other than the nuclear issue. However, if the dialogue between the superpowers were to develop and cover regional problems, keeping the "bomb" issue completely out of the picture may be difficult, and the prospect of a Middle Eastern nuclear-free zone – occasionally discussed – may become a less remote prospect, and even in the interest of Israeli security. Tel Aviv and Moscow have been working for quite some time now on re-establishing diplomatic relations. Perhaps Fidel Castro can be forced to go down the same path with Washington.

Concluding Remarks and Queries

The four major developments on the international scene that were listed before may substantially change that scene by 1995 – the year of the NPT renewal – from the way it is today. Excessive extrapolation should of course be avoided. The process of change is subject to slow-downs here and there, and possibly also to reversals. A high degree of unpredictability must be assumed.

Two principal sets of questions are likely to arise concerning the future of the non-proliferation regime if we project our analysis, indeed our speculation, five or six years ahead in this state of high flux. One is about the evolution in some of the countries of Eastern Europe that may lead to greater autonomy, or possibly even the neutrality or semi-neutrality of one or two. Would Moscow's nuclear guarantee remain and would this guarantee still coincide with the sphere of security of the Soviet Union, even if its sphere of influence were modified by a shift towards the West in the economic and possibly the socio-cultural life of these countries? Will this eventual "Finlandisation" of some Warsaw Pact members affect the nuclear non-proliferation regime before 1995?

Secondly, the shrinking Soviet political – if not strategic – hegemony may be moving parallel to a confirmed decline of American economic – if not political – hegemony. Will the strategic nuclear balance nevertheless remain strong and extended enough not only to maintain deterrence but also to continue to discourage attempts to build regional nuclear balances of dubious stability? Moreover, the dual leadership of the superpowers has been the main engine in advancing the non-proliferation regime and containing, if not solving, the remaining "problem situations" that affect and limit the regime itself. If this dual leadership is weakened, will multilateral cooperation be able to supplement it and be effective enough to bring about a renewal, possibly an improvement, of the NPT in 1995?

References

1 See, for example, J. J. Weltman, "Nuclear Devolution and World Order" in *World Politics*, (January 1980) pp. 169–93; and K. Waltz, "The Spread of Nuclear Weapons: More May be Better" *Adelphi Paper*, No. 171, (London: IISS, 1981). The porcupine image is used by R. Sandoval, "Consider the Porcupine: Another View of Nuclear Proliferation", in *Bulletin Atomic Scientists*, (May 1976) p. 17

2 See S. Feldman, *Israeli Nuclear Deterrence. A Strategy for the 1980s*, (New York: Columbia University Press, 1982). Feldman is a pupil of K. Waltz

3 See Edward Teller, "The Feasibility of Arms Control and the Principle of Openness" in D. G. Brennan (ed.), *Arms Control, Disarmament and National Security*, (G. Braziller, 1961) p. 134

4 The word "unnatural" is used, for instance, by J. Nye, "Prospects for Non-Proliferation" in R. W. Jones, C. Merlini, J. F. Pilat, and W. C. Potter, *The Nuclear Supplies and Non-Proliferation*, (Lexington, MA: Lexington Books, 1985) p. 219

5 "What the President Saw", interview with R. Nixon, *Time*, (29 July 1985) p. 43

6 Interview with H. Kissinger, *La Stampa*, 11 August 1985, p. 5

7 Because of their technological capabilities, these countries can be called "potential problem countries". I have developed the point in C. Merlini, "Problem Countries" in R. W. Jones, C. Merlini, J. F. Pilat, and W. C. Potter, *op. cit.*, p. 155

8 For an analysis of control of nuclear weapons in Western Europe, see P. Cotta Ramusino, *Il controllo delle armi nucleari in Italia*, (Rome: CESPI, April 1985). I also received useful suggestions from M. Carnovale and M. De Andreis, IAI

9 *High Frontier* is the rightwing political movement that became famous because it came up with the SDI proposal which Reagan endorsed. See D. O. Graham, Project Director, High Frontier, *We Must Defend America: A Strategy for National Survival*, (Chicago: Regnery Gateway, 1983)

10 See C. Merlini, Audizione alla Commissione III (Affari Esteri) dalla Camera dei Deputati, IX Legislatura, Indagini conoscitivi e documentazioni legislative, Seduta antimeridiana di mercoledi; 9 ottobre 1985

11 See G. Allison, A. Carnasale, and J. Nye, "The Owls' Agenda for Avoiding Nuclear War" in *The Washington Quarterly*, (Summer 1986)

12 See S. Nunn and S. W. Warner, "Reducing the Risk of Nuclear War" in *The Washingtion Quarterly*, (Spring 1984) p. 3

13 See H. Bennendijk, "NATO's Nuclear Modernization Dilemma", *Survival* 3/4, (London: IISS, 1989)

14 Bettino Craxi, leader of the Italian Socialist Party, then also Prime Minister

15 H. Kissinger, "East Asia, the Pacific and the West: Strategic Trends and Implications", *Adelphi Paper*, No. 216 (London: IISS)

16 The member countries of the Association of South-East Asian Nations (Brunei, Indonesia, Malaysia, the Philippines, Singapore and Thailand) have declared their territories programmatically free of nuclear weapons

Part IV

Priorities for Pursuing the Objectives of Article VI of the NPT

Seismic Monitoring of a Multilateral Nuclear Test Ban Treaty

Jonathan Berger

1. Introduction

This paper discusses the role of seismic monitoring in the verification of a multilateral nuclear test ban treaty. The current capabilites of the seismic method of detecting and identifying nuclear explosions are reviewed. The earth's natural seismic activity is described, as it is in this environment that a global monitoring system must operate. The current state of global seismic monitoring is decribed. Finally, a practical proposal is presented for an international global seismic monitoring system that would provide, beyond the role of monitoring treaty compliance, considerable social and scientific benefit.

The principal tasks of seismic monitoring are:

- Detection and location of all seismic events larger than a certain size.
- The identification of detected events, that is, the discrimination of clandestine nuclear tests from earthquakes and industrial explosions.

As we shall see, these two tasks can only be performed imperfectly; there will always be a margin of error and uncertainty in the results. As the magnitude of the event gets smaller, there are

- more earthquakes and industrial explosions,
- more possibilities for treaty evasion, and
- more factors that influence the seismic signal produced.

For intermediate and large yield tests, the seismic method is clearly capable of identifying the source with little ambiguity. For small yields, however, the seismic monitoring method will only help to identify events that are of a suspicious nature; that is events which might be clandestine nuclear tests. Studies in the USA and the USSR indicated that with a network of high-performance stations, seismic monitoring down to the tamped 5 kiloton (kt) yield would be practical with a high degree of certainty. Thus, while the seismic method of monitoring a threshold test limitation treaty could be quite accurate and complete, a comprehensive test ban treaty can at best be monitored by these methods only down to a level of yield that makes clandestine weapons development very difficult. Below some level of yield, uncertainties of the seismic method are sufficiently great that other techniques of monitoring, such as on-site inspection, will have to be negotiated. Even then, the seismic method can still be used to trigger on-site inspections and guide inspection teams to the source region. Further, there are considerable societal benefits that would result from the development of such a global monitoring network beyond the arms control issues. Examples of these benefits are rapid earthquake location and magnitude determination for damage assessment and emergency services, tsunami warning, possible application to earthquake warning, and the provision of a data base for scientific studies of the earth's structure and earthquake mechanisms.

2. Earth's Seismicity

The earth's naturally occurring seismic activity or seismicity, that is, the patterns of earthquake activity that normally occur, forms the "environment" in which a seismographic monitoring system must operate. Each seismic event larger than a certain magnitude must be detected and examined to determine if it was an earthquake, an industrial explosion, or perhaps a suspected clandestine nuclear test.

A fundemental observed property of the earth's seismicity is that the number of events larger than a certain magnitude increases as the magnitude decreases. Mathematically, this relation can be expressed approximately by the formula

$$\log N_Q = a - bm_b \quad ,$$

where N_Q is the number of events of magnitude m_b or larger, and a and b are constants. For the entire world, a and b are approximately 7.4 and 0.9, respectively. There are about 1000 events annually with magnitude larger than 5; about 60,000 earthquakes per year (roughly one every ten minutes worldwide with magnitudes greater than 3; and nearly 200,000 events annually with magnitudes larger than 2.5!

The second important observed feature of the world's seismicity is that most earthquakes occur in well delineated seismic zones, as is clearly illustrated in Figure 1. The majority of activity, in fact, occurs in oceanic regions where the major tectonic plates abut. For example, about two-thirds of the seismicity in and surrounding the Soviet Union occurs in the oceanic areas off the Kurile-Kamchatka coast, at the boundary of the Pacific and North American plate. About 5 % of all earthquakes occur at depths below 300 km. At intermediate depths of between 60 and 300 km, about 25 % occur. The remainting 70 % occur in the earth's uppermost crust at depths less than 60 km. Almost no earthquakes, however, occur at depths as shallow as those used for nuclear testing.

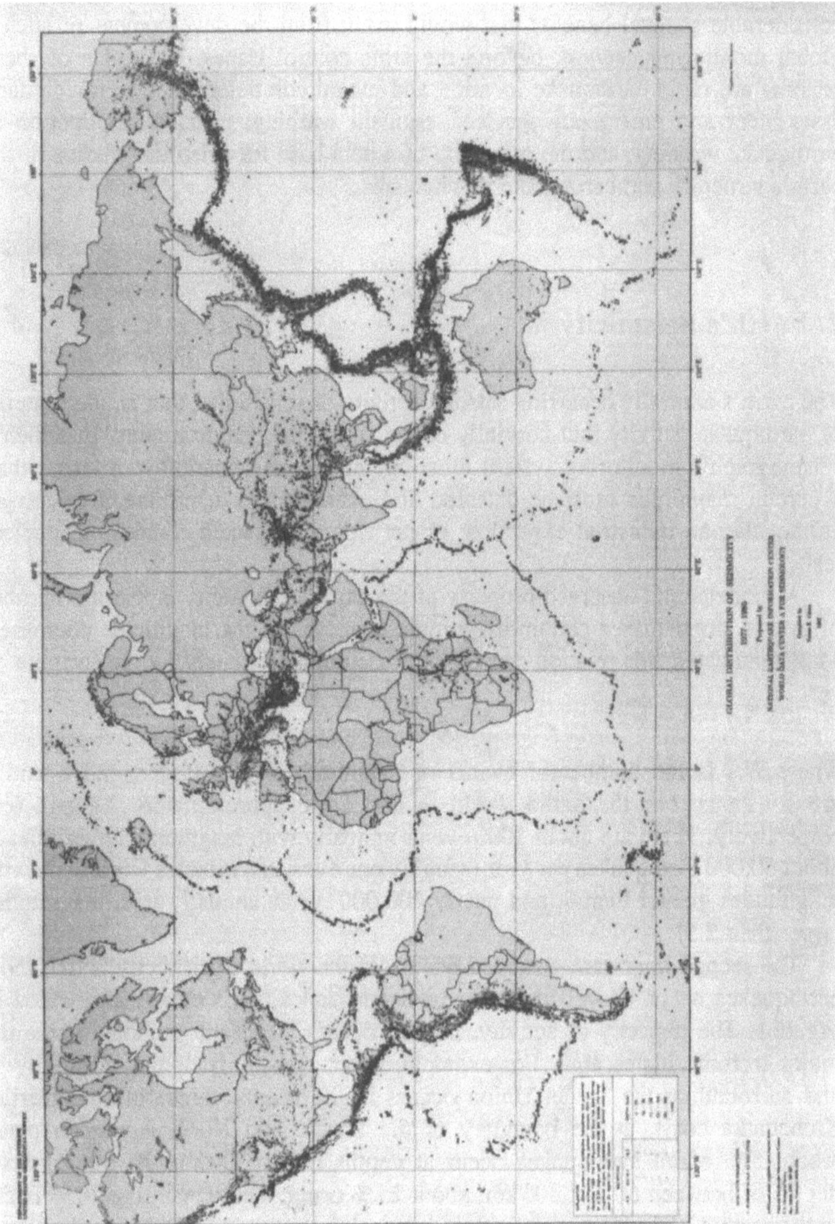

3. Yield Versus Magnitude

To compare earthquakes and explosions it is convenient to express the yield of an explosion in terms of an equivalent earthquake magnitude, which itself is an estimate of the energy release. Much work has been done on this problem, chiefly by observing nuclear and chemical tests of known yields with seismic instrumentation and calculating the magnitude of these tests in the normal manner.

Roughly speaking, a 1 kiloton yield is equivalent to a magnitude 4 if the detonation is "well-coupled" and in hard rock. Well-coupled means that the energy transfer from the explosion to the surrounding rock is efficient. The geology and other conditions at the shot point affect the coupling, however, and the regional geology affects the transmission of the seismic waves from the explosion to the observation point as illustrated in Figure 2. These effects lead to a different relation between yield and magnitude for the US Nevada Test Site and the Soviet Kazakh Test Site, the so called "magnitude bias", estimated at about 0.3 to 0.4 magnitude units. Tests in Kazakhstan are observed to be about this much larger in magnitude than the equivalent yield test in Nevada. Overall, these variations can lead to an uncertainty of up to a factor of 10 in the yield for a given magnitude. Even larger variations can be caused by "de-coupling" or deliberately reducing the transfer of energy from the blast to the surrounding rocks. This can be done by detonating the test in a large underground cavity. Decoupling can

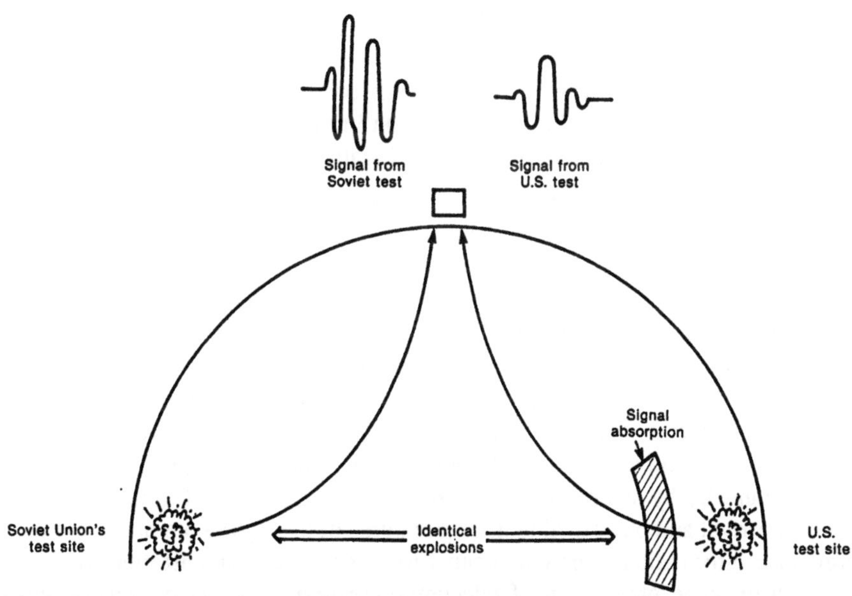

Fig. 2. Schematic illustration of attenuation-related magnitude bias. Seismic body waves crossing parts of the upper mantle with high temperatures become reduced in amplitude. Seismic signals from the Soviet Union's test site appear much larger than signals from the same yield explosion conducted at the US test site. (Source: Congress of the United States, Office of Technology Assessment.)

reduce the apparent magnitude by up to a factor of 70 so that a 7 kt test appears to be a 0.1 kt test. There are practical limits to this method, however, as the size of the cavity required is quite large for even a modest yield. It is estimated that naturally occurring cavities limit this method of evasion to yields of about 10 kt or less. Table 1 summarizes the relation of magnitude to yield for well-coupled explosions in hard rock (the usual test conditions), and the earth's normal seismicity.

Table 1. Magnitude versus yield (fully-coupled in hard rock) and number of events globally of equal or larger magnitude

Magnitude	Yield (kilotons)	No of Events/year
3	0.1	60,000
3.5	0.5	20,000
4	1	7,500
4.5	5	2,700

4. Methods of Detection and Location

Due to the structure of the earth's interior, the paths taken by the seismic waves, as they radiate away from their source, are complicated and depend upon the distance from the source to the receiver. In general, however, the further a station is from the source the smaller will be the signals from a given size event. This is the result of:

- the spreading of the wavefront as it expands outwards from the source;
- the dissipation of energy through absorption in the earth; and
- energy refraction, diffraction, reflection and scattering at subsurface boundaries.

Modern instrumentation, however, can detect any signal in a frequency band whose energy exceeds that of the ambient background earth noise. Further, techniques of noise suppression or signal enhancement have been developed and are widely employed. Stations equipped with modern instrumentation and located in a low-noise environment will usually be able to detect signals from events of magnitude 4 or larger anywhere within about 5000 km of the station.

To locate an event with a reasonable degree of accuracy requires the detection of signals at several stations. Location of the event is straightforward provided that signals are received by at least three stations. Traditionally, signals are "detected" manually by station personnel from analogue records of ground motion. Times of arrival and other distinguishing features of the signals are noted and this parametric information is communicated to some central organization by mail, telegraph or some other form of telecommunications. The set of stations thus reporting is what is commonly called a *seismographic network*. Based on this set of parametric measurements, the central organization determines the location and magnitude of the event by straightforward processing. Depending upon the form of communication used by the network stations, the location and magnitude might take weeks, or even months to determine.

Today, most modern networks rely on some form of rapid reporting to locate and determine magnitudes within a few days. It is, however, rare that the signals themselves are reported to the network centre. Yet, it is precisely these signals that must be scrutinized in detail to provide a more detailed estimation of the nature of the source. In a treaty monitoring regime, such analysis would be mandatory.

With an adequate number of high-performance stations, detection of seismic signals from events of magnitude 3 and larger is not difficult. The more stations that detect the signal, the more accurate the location. Typical uncertainties are 5 km on the surface projection and 30 km in the depth.

5. Methods of Identification

How can one tell an explosion from an earthquake? How can one tell a nuclear explosion from and industrial explosion? The principal discriminants between earthquakes and explosions are:

Location. This is the principal method of identifying an event as an earthquake. If the location is clearly in an oceanic and seismically active area, the presumption is that the event is an earthquake. The absence of a hydro-acoustical signal associated with an event located in an oceanic area is strong, if not conclusive, evidence of natural origin. If the event is located on land, it may still be considered a natural event if the location is in either an earthquake-prone area or in an unlikly spot for a nuclear test. Conversely, if an event occurs in an aseismic area where testing might occur, suspicions will be raised.

Depth. The depths of seismic events can be determined fairly accurately if the event is well-recorded and the depth is greater than about 15 km. Such events can be classified as natural with a high degree of confidence. For those occurring in the upper 15 km of the crust, depth determination is difficult and prone to large errors, although this is an area of active research in which improvements can be expected.

M_s/m_b. This discriminant is based on the ratio of magnitudes determined by surface waves which, as their name implies, travel along the surface of the earth, and body waves which travel deep within the earth. Due to the difference between the mechanisms of a nuclear explosion and an earthquake, this ratio usually is different for these two classes of sources. The earthquake ruptures over hundreds of metres to kilometres, and the rupture travels at about 3 km/sec. Thus, the time of generation of the earthquake waves is typically fractions of a second to several seconds. A nuclear explosion releases its energy in microseconds. The M_s/m_b discriminant works fairly well for events which produce both kinds of signals with amplitudes large enough to measure at a suite of stations, but for events of magnitude smaller than about 4.5 it is unreliable.

Other Methods. There are many other methods of identification which rely on more sophisticated analysis of the signals produced by the seismic source. These are a subject of active research by the seismological community and may prove to be helpful in the identification of events of suspicious origin. By themselves, they are unlikely to provide definitive results unless some breakthrough in research occurs.

To distinguish nuclear from chemical explosions on the basis of seismological data alone is more difficult than to distinguish earthquakes from explosions. The chemical explosions that are used routinely for mining and construction purposes, however, do have some source features that are different from nuclear explosions. These arise from the way in which the chemical explosions of this sort are detonated. Most industrial explosions (chemical) have yields of 0.1 kt or less, and are, in fact, detonated as a series of many small explosions separated in space by a few metres, and in time by tens of milliseconds. These so called "ripple"-fired blasts radiate seismic signals whose spectral properties are significantly different from both earthquakes and explosions which are detonated at one place and time, so called "salvo"-fired blasts. Frequent periodicities in the spectrum of the radiated seismic signals is a clear indication of a ripple fired blast.

6. Estimates of Performance of Seismic Monitoring

The Soviet Union and the United States are, without doubt, the most studied regions in terms of seismic monitoring scenarios. Further, as they the most active in nuclear weapons development and testing, and their territories are large, the challenge of monitoring a test ban treaty is perhaps the greatest in these areas.

In the Soviet Union, the principal nuclear weapons test sites are in eastern Kazakhstan and in Novaya Zemlya, both aseismic areas. Thus, the Soviet Union can serve as a model of monitoring other aseismic areas of the world where testing might be expected. Examples are Argentina, Israel, and South Africa. Until recently, however, certain essential data that could only be obtained by direct seismographic measurements were not available from Soviet territory. Then, in 1986, the Natural Resources Defense Council and the Soviet Academy of Sciences concluded an agreement under which three seismic stations were established in the vicinity of the principal nuclear-weapon test sites in eastern Kazakhstan and Nevada. Not only did this landmark project demonstrate the willingness of the Soviets to allow US monitoring stations on their territory, but data from these stations have led to improved estimates of the capabilities of a network of such in-country stations to monitor future test ban treaties. During the period of operation of the NRDC stations around the Kazakh test site, virtually all the recorded seismicity in the area was at locations known to be areas of mining. No seismic activity of clearly natural origin was recorded from the test site.

In marked contrast, the US Nevada Test Site is located in the active Basin and Range tectonic province, where naturally occurring seismic activity is prevalent. Studies of the monitoring of this test site are good models for seismic areas as are found in, for example, Pakistan, India, China, Iraq, and Iran. The overall seismicity for the USA and USSR territories is, however, similar in quantity and certain parallels regarding seismic monitoring can be made.

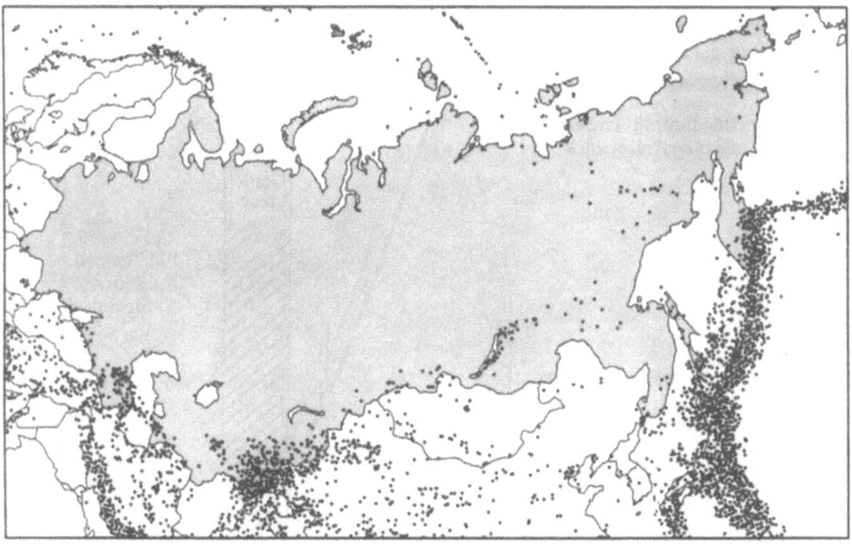

Fig. 3. Seismicity of the USSR and surrounding areas for the 15 year period 1971–1986. (Source: United States Geological Survey)

The seismicity in and around the Soviet Union is illustrated in Figure 3. As is clearly evident in this figure, most of the seismic activity occurs along the southern and eastern borders of the country. Along the eastern borders, moreover, most of the earthquakes occur offshore in the Kurile-Kamchatka region. Here, the shallow earthquakes occur offshore while the deeper events occur under the land areas. For a comparison with the seismicity of the USA see Figure 1, which illustrates approximate parity when Alaska (particularly the Aleutian arc) is taken into account.

Figure 4 graphically illustrates estimates of the number of unidentified events per year in the US/USSR as a function of magnitude (and decoupled yield). At the 5 kt decoupled yield, it is estimated that up to 10 events per year would be unidentified and would require some other kind of verification, such as on-site inspection. Note, however, that the seismic network would detect and locate all these events and would thus serve as a trigger for on-site inspection.

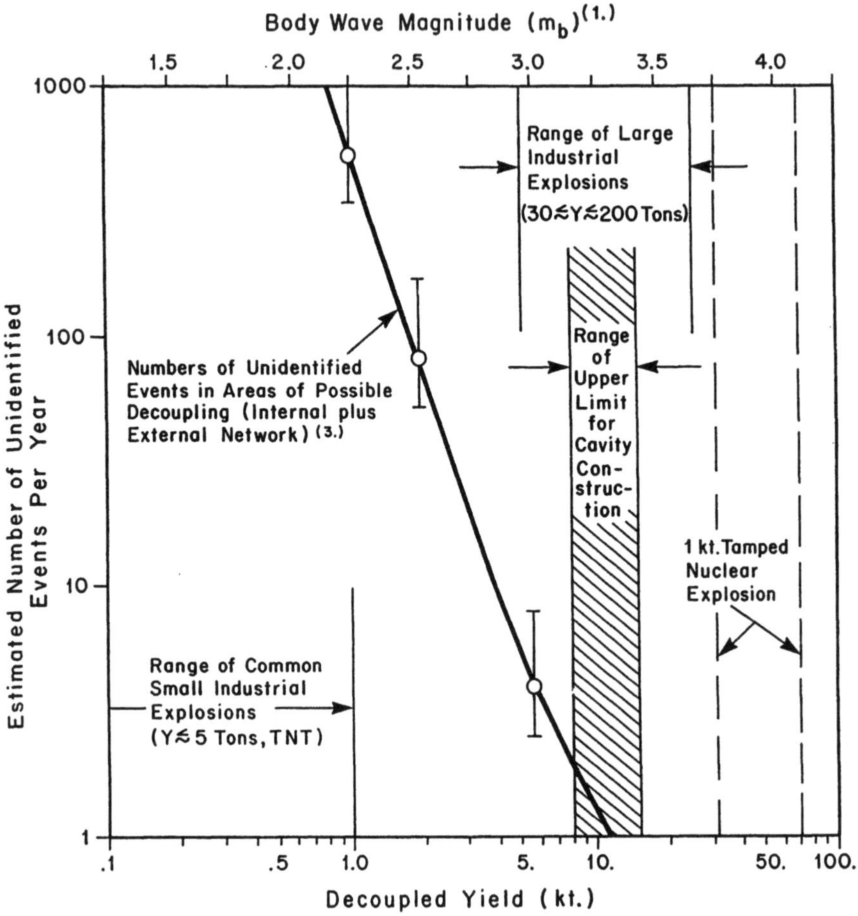

Fig. 4. Estimated cumulative numbers of unidentified seismic events in the USA/USSR per year versus decoupled yield or magnitude, showing range of small and large industrial explosions, upper limits for cavity decoupling and magnitude of 1 kt well-coupled explosions

7. Principal Problems with Seismic Monitoring

Operational Difficulties. The chief operational difficulty in the routine operation of a monitoring network would be the occasional overloading of the system due to swarms of earthquakes or large numbers of industrial explosions. Large earthquakes are often followed by swarms of smaller events, sometimes so frequent that the signals from one event overlap the signals from another, and the monitoring system is presented with nearly continuous signals for some period of time. Fortunately, such occurrences are rare, and the area affected is usually of fairly small dimensions.

Certain areas, such as mining or quarrying areas, will usually be the source of many small industrial explosions. Most of these can be identified by location, particularly if, by treaty agreement, those areas and routine patterns of detonation are specified.

Another kind of overload can occur when events are suspicious and require in-depth analysis. Problematical events can easily take in excess of a day to process and analyse. It is estimated that for a region the size and seismicity of the USSR, about 30 events per day would have to be processed with perhaps 20 % of these requiring analysis beyond simply location and depth determination.

Detection Evasion. Various evasion schemes have been proposed during the course of the years of test limitation treaty negotiations. How practical they are is a matter of debate, but the net effect on a seismic monitoring system is that more processing is required to defeat, or at least make more difficult, such evasion techniques.

Detonating a test in a large underground cavity makes the seismic signals smaller by up to a factor of 70. Thus, to assure compliance down to some yield level, events of about 1.5 magnitude units lower must be processed. The number occurring naturally at this lower magnitude is increased. However, this particular evasion technique is only considered viable in certain areas where such underground cavities occur naturally, and thus the extent of such heightened seismic surveillance would be limited. (The construction of large cavities would be visible to reconnaissance satellites or detectable by other means.)

Other evasion techniques that have been suggested as credible include the detonation of a decoupled nuclear test at the time of a large nearby industrial explosion. Seismic means would proably not be able to resolve this, and some treaty agreement for on-site inspection/monitoring of large industrial explosions, in areas where decoupling cavities might naturally exist, would need to be negotiated.

False Alarms. Estimates of the false alarm rate for the USA/USSR imply that about 5 to 10 events per year with magnitudes above 3 would ultimately be either unidentified or mis-identified provided that large industrial explosions are announced. This is the estimated current limit on the seismic method, and presumably on-site inspections or some other verification method would be required.

8. An International Seismic Monitoring System

The first standardized worldwide seismographic network came about as a result of the test ban discussions held in 1958. The United States began Project Vela Uniform managed by the Defense Advanced Research Projects Agency in response to a government study referred to as the "Berkener Report" named after the panel chairman. One of the recommendations of this report led to the establishment of the World-Wide Standardized Seismographic Network (WWSSN).

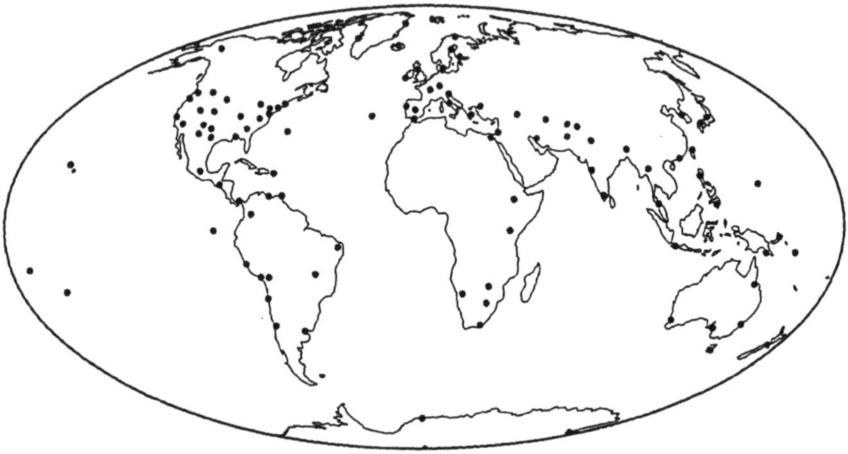

Fig. 5. World-Wide Standardized Seismographic Network (WWSSN) station distribution. (Source: United States Geological Survey.)

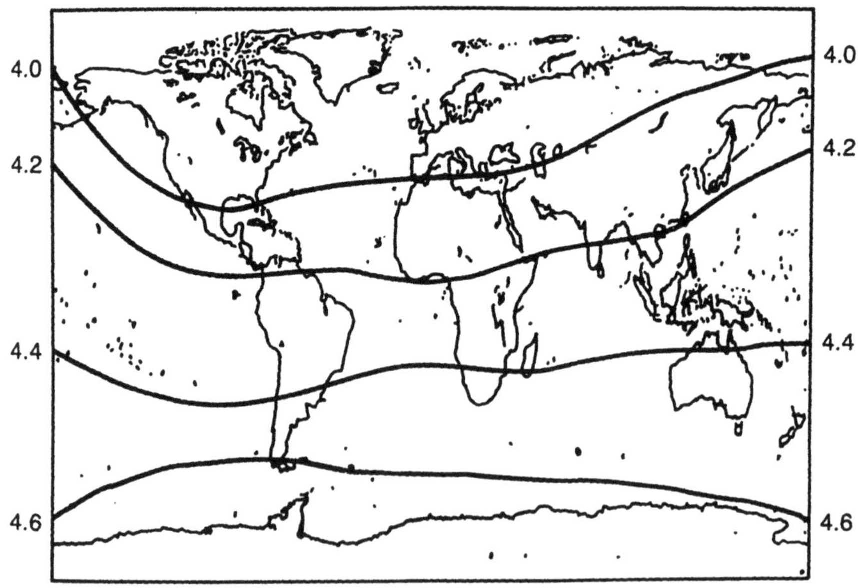

Fig. 6. Countours of 90% probability of detection at at least 4 stations of the present international global network. (Source: US Defense Advanced Research Projects Agency.)

The endeavour was an outstanding success from several points of view. First, it was a truly international scientific project. At its peak in 1967, the network consisted of some 120 stations installed in 50 countries (See Figure 5). Second, the quality of the data and the data distribution system led to a revolution in the science of seismology and tectonics. Our current ideas of plate tectonics and the formation of the continents were developed in large part as the result of data from this network. Finally, it fulfilled its principal purpose in detecting and locating nearly all the nuclear tests over a certain size that were conducted underground. Data from the network firmly established the seismic method of nuclear test detection, identification and yield determination in the Limited Test Ban Treaty regime.

Today, however, the network is obsolete and operating at about half its peak level in terms of the number of stations routinely reporting data. More importantly, the instrumentation, the data collection system, and the data recording medium, are all clearly inadequate for the task of monitoring a threshold or comprehensive test ban treaty. The data are recorded in an analogue manner by photographic or other graphic means. At some stations the records are daily analysed by hand with significant results telegraphed to the data centre operated by the United States Geological Survey in Boulder, Colorado. The actual graphic records are sent by some stations to the data centre on request by mail, and complete analysis cannot be done for months.

Figure 6 illustrates the actual detection performance of the existing informal international network which, besides the WWSSN stations, consists of stations run by numerous other organizations. The degrading of performance in southern latitudes is a result of very sparse coverage in Africa and South America. In marked contrast, Figure 7 shows the projected detectability for a monitoring network of high-performance stations in and around the USSR. Note the considerably poorer detectability of the actual network shown in this figure compared with that calculated for the modern high-performance network.

In the past 30 years since the WWSSN was designed, seismic instrumentation has made huge advances, as modern electronics and communications have been

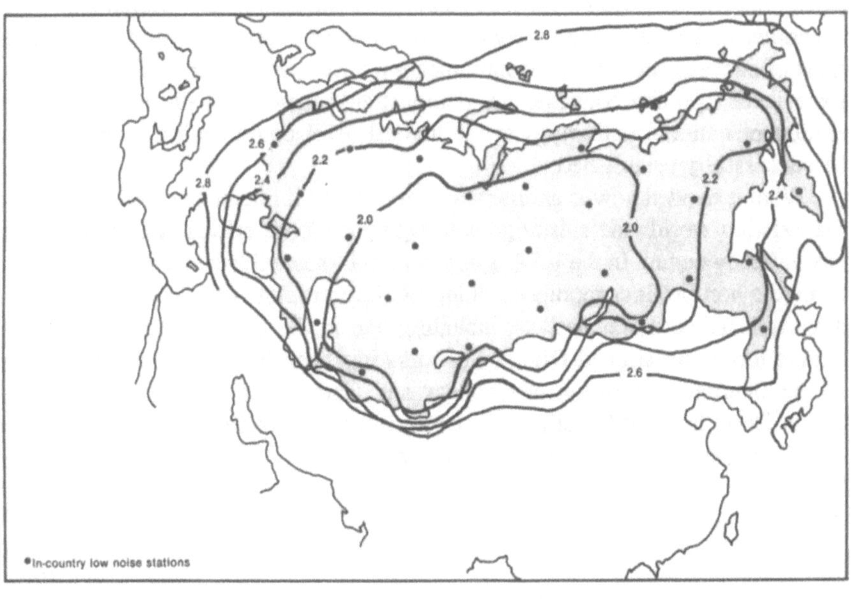

Fig. 7. Estimated detection capability of a network of modern high-performance stations (Source: Congress of the United States, Office of Technology Assessment.)

applied. The technology has been developed and tested under field conditions (such as the NRDC project in the area of the Kazakh and Nevada Test Sites). The key features of such a global seismographic network would be:

- approximately 100 seismic monitoring stations deployed worldwide;
- extra stations located in areas of intensified monitoring;
- data telemetered in near-real time to international data centres;
- rapid data processing to detect, locate and identify all seismic events of magnitude larger than about 3.5 worldwide;
- enhanced network capabilities for certain geographical areas of particular interest;
- provision of rapid earthquake locations and tsunami warning, with applications to earthquake warning, earthquake hazard assessment, and prediction;
- provision of a global data base for scientific studies of the earth's structure, earthquake source properties, and other geophysical problems.

The key element in the implementation of such a global seismic monitoring system is international cooperation both for the installation of stations in many countries and for *communications*. Without the rapid collection and processing of data from a seismographic network, the results would not be useful in the monitoring regime, and be of little other societal value. International communications via satellite and other technologies is, of course, commonplace today. But nonetheless, it is highly controlled by national authorities under a plethora of regulations and tariffs. It is fairly easy to establish a seismic station within a given region by working within existing scientific organizations. Establishing real-time communications between that station and some international data centre is another matter, involving a completely different bureaucracy and raising questions of national security, telecommuncation tariffs, and other arcane regulations. Experience in many countries has demonstrated that only with the active cooperation of national authorities at the highest level could such a global network be successfully established.

What is needed now is an international effort to implement a modernization of the existing worldwide siesmographic network. Some efforts at implementation have already begun. In the USA, a consortium of some 57 research institutions has formed a non-profit corporation, called IRIS, to promote this programme and has already installed some stations, including five in the USSR. Internationally, the Federation of Broadband Digital Networks has been formed within the auspices of the International Union of Geodesy and Geophysics to work towards these goals. Costs have been estimated at 50 million dollars for capital expenses and 10 million per year for operations and maintenance.

9. Conclusions

1. There is general agreement that with seismic monitoring, a 5 kt to 10 kt threshold test ban treaty is verifiable.
2. The technical and seismological expertise exists *today* to implement such a monitoring network.
3. The most active countries in nuclear weapons testing, the USA and the USSR, are the most difficult to monitor due to their size and natural seismicity. They have both, however, demonstrated their willingness to accept seismographic stations on their territories for monitoring purposes.
4. An international seismographic network that could provide the basis for any multilateral test ban treaty monitoring would, in addition, also provide a much needed earthquake monitoring system for rapid earthquake hazard assessment, tsunami warning, and scientific research.

The Linkage Between a Nuclear Test Ban and Nuclear Non-Proliferation

William Epstein

I. Legal Perspective

As the years go by, there is an increasing tendency by those who do not favour a CTB to ignore and, some observers believe, to suppress and distort, or engage in an *ex post facto* re-interpretation of the meaning and intent of the NPT, in order to deny or weaken the link between a CTB and the NPT. A new generation has come into existence since the conclusion of the PTBT in 1963 and the NPT in 1968 that had no direct experience of the complicated developments. As one who was involved in the negotiations leading up to the two treaties, I feel it is necessary to review briefly the "legislative" history of the treaties in order to keep their interrelationship clearly in mind.

The negotiations in the tripartite Conference on the Discontinuance of Nuclear Tests and in the ENDC were all devoted to the achievement of a total test ban treaty.

After the Cuban Missile Crisis of October 1962, both the USA and the USSR were anxious to make progress on arms control. Premier Nikita Khrushchev agreed to allow automatic seismic stations to be placed in the Soviet Union and to permit two or three on-site inspections annually on Soviet territory to monitor underground testing. President John F. Kennedy insisted on eight (later reduced to seven) annual inspections. The non-aligned members of the Geneva Disarmament Committee proposed a compromise of four or five on-site inspections, but to no avail. Glenn T. Seaborg, then the head of the US Atomic Energy Commission, and who attended the signing of Partial Ban Treaty in Moscow, has written, "I regard the failure to achieve a comprehensive test ban as a world tragedy of the first magnitude". I can heartily endorse that statement. I regard it as one of the great lost opportunities of history that stubborn elements within each of the two superpowers could block the achievement of a reasonable and attainable CTB agreement.

The preamble of the final agreed text of the Partial Test Ban Treaty states:

> "Seeking to achieve the discontinuance of all test explosions of nuclear weapons for all time, determined to continue negotiations to that end, and desiring to put an end to the contamination of man's environment by radioactive substances."

Article I (b) refers to:

"... the conclusion of a treaty resulting in the permanent banning of all nuclear test explosions, including all such explosions underground, the conclusion of which, as the Parties have stated in the Preamble of this Treaty, they seek to achieve."

Article II sets out the amendment procedure. Article III provides for the signature or accession by all states. Mr. Seaborg notes:

"From the outset both sides had agreed on the importance of having a maximum number of nations adhere to the treaty. In this way it was felt that the treaty might achieve one of its primary objectives, to inhibit the further proliferation of nuclear weapons capability" [1].

The above-mentioned provisions provide clear evidence that the treaty was intended not merely as a step towards ending the armaments race and as an environmental measure but also, although not explicitly so stated, as a non-proliferation measure [2]. As indicated earlier, however, the failure to achieve the end of all testing has kept it from being transformed into a really important measure of arms control or non-proliferation.

It is noteworthy that when the General Assembly considered and approved the 1963 PTBT it singled out and emphasized the provisions of both the preamble and operative parts of the Treaty referring to a total ban on testing. In this connection, the resolution (1910 XVIII) states:

"Noting further with satisfaction that in the preamble ... the parties State that they are seeking to achieve the discontinuance of all test explosions of nuclear weapons and are determined to continue negotiations to this end ...

2. Requests the Eighteen-Nation Committee on Disarmament to continue with a sense of urgency its negotiations to achieve the objectives set forth in the preamble of the Treaty.
3. Requests the Eighteen-Nation Committee to report to the General Assembly at the earliest possible date and, in any event, not later than the [next] session."

This action would seem to make it clear that the members of the General Assembly expected rapid progress towards a CTB. This expectation by the Assembly was confirmed by a long series of subsequent annual resolutions regretting the failure to achieve a CTB, condemning continued testing, calling for a suspension of all nuclear weapon tests, and urging completion of a CTB as a matter of highest priority.

When the NPT was being actively discussed and promoted in the UN General Assembly, the eight non-aligned members of the ENDC (Brazil, Burma, Egypt, Ethiopia, India, Mexico, Nigeria and Sweden) proposed a resolution in 1965 calling on the ENDC to negotiate the NPT based on the following main principles:

a) the treaty should be void of any loop-holes which might permit nuclear or non-nuclear powers to proliferate, directly or indirectly, nuclear weapons in any form;
b) the treaty should embody an acceptable balance of mutual responsibilities and obligations of the nuclear and non-nuclear powers;

c) the treaty should be a step towards the achievement of general and complete disarmament and, more particularly, nuclear disarmament;
d) there should be acceptable and workable provisions to ensure the effectiveness of the treaty;
e) nothing in the treaty should adversely affect the right of any group of States to conclude regional treaties in order to ensure the total absence of nuclear weapons in their respective territories.

This resolution, which become famous as "Resolution 2028", thereafter became the basis of the positions of the Non-Aligned countries in the ENDC, who put primary emphasis on the necessity for any non-proliferation treaty for an acceptable balance of mutual responsibilities and obligations on the part of the nuclear and non-nuclear powers.

It was, moreover, the insistence of the non-aligned members that led to the inclusion in the preamble of the 9th, 11th and 12th paragraphs concerning the cessation of the nuclear arms race, the achievement of a CTB and the elimination of nuclear weapons.

Article VI embodies what is in effect a promise by the three nuclear powers to end their vertical proliferation of nuclear weapons in exchange for the clear commitment of the non-nuclear states not to engage in the horizontal proliferation of these weapons. Earlier, the eight non-aligned members had submitted a joint memorandum which put primary emphasis on the necessity for an acceptable balance of mutual responsibilities and obligations between the nuclear and the non-nuclear states, in accordance with General Assembly resolution 2028 adopted in 1965. Among the measures proposed in the memorandum (in the order listed) were:

- a comprehensive test ban,
- a complete cessation of the production of fissionable material for weapons purposes,
- a freeze on, and a gradual reduction of, nuclear weapons stocks and means of delivery,
- a ban on the use of nuclear weapons, and
- security assurances to the non-nuclear states.

The non-aligned members bargained hard for the inclusion of these measures and were supported by Canada, Bulgaria and Romania. When the three nuclear powers maintained that it would be counter-productive to list specific measures in the operative part of the treaty, the non-nuclear countries tried various compromise formulations, with a CTB as the first measure in each case, but to no avail. The most that the three nuclear powers would concede on this matter was the inclusion of paragraphs 9, 11 and 12 in the preamble, and the inclusion of the italicized phrases "pursue negotiations *in good faith*" for the cessation of the nuclear arms race "*at an early date*" in Article VI [3].

The final agreed text of Article VI reads:

"Each of the Parties to the Treaty undertakes to pursue negotiations in good faith on effective measures relating to cessation of the nuclear arms race at an early date and to nuclear disarmament, and on a treaty on general and complete disarmament under strict and effective international control."

In addition, the nuclear powers agreed in Article VIII to make provision for amendment conferences at the request of one-third of the parties, and for review conferences. In Article X they agreed to a right of withdrawal of any party and, after 25 years, to a conference to decide on the renewal of the Treaty.

As regards the interpretation of Article VI, it was well understood by all parties that "measures relating to cessation of the nuclear arms race" was clearly intended to cover such measures as a CTB, a ban on the production of fissionable material for weapons, a freeze on the production of additional nuclear weapons, and a ban on flight testing of delivery vehicles. While there was room for differences regarding the entire list, a CTB was at the top of everyone's list of measures, and there can be no doubt that a CTB was considered as the essential and indispensable measure for a cessation of the nuclear arms race.

As in the case of the PTBT, it is noteworthy that the UN General Assembly emphasized the importance of Article VI of the NPT. UN resolution 2373 (XXII), adopted by 95–4–21 on 12 June 1968, commended the NPT and expressed the hope for the widest possible adherence by both nuclear-weapon states and non-nuclear-weapon states. This time, the resolution singled out Article VI of the Treaty for special treatment. Operative paragraph 4 of the resolution reads:

"Requests the Conference of the Eighteen-Nation Committee and the nuclear-weapon states urgently to pursue negotiations on effective measures relating to the cessation of the nuclear arms race at an early date and to nuclear disarmament and on a treaty on general and complete disarmament under strict and effective international control."

The quotation of the substance of Article VI of the Treaty in the operative part of the UN resolution would seem to make it clear that the community of nations, as well as the nuclear powers, recognized that this commitment was the most important one undertaken by the parties.

Although it could in logic and reason be argued that a CTB might not by itself be sufficient to ensure a cessation of the nuclear arms race, it is clear that there could be no cessation without a CTB.

The "cessation of the nuclear arms race at an early date" precedes "nuclear disarmament" in the text, which gives it priority according to standard rules of legal interpretation [4], as indeed it does in reason and logic. As the Canadian member of the ENDC pointed out at the time, in order for the nuclear arms race to be reversed, as in the case of a motor vehicle or ship, it would first have to be halted or brought to a stop. Hence the cutting back of nuclear weapons must follow rather than precede a CTB. This line of accepted interpretation and reasoning was for the first time challenged by the Reagan Administration and a few of its allies, which have insisted that "deep cuts" deserve the highest priority and that a CTB can be relegated to a "long term" goal as a part of the "ultimate" process of the elimination of nuclear weapons. It is submitted that this is not only a misinterpretation or re-interpretation of Article VI, but that it is an untenable proposition on its merits.

It seems obvious, and it has been pointed out on a number of occasions, that large reductions as well as small ones in the nuclear arsenals of the nuclear powers could be nullified or become meaningless if these powers continue to test new or modernized nuclear weapons which would be more accurate, dangerous and destabilizing than those they replace. This could lead to a situation where "less is more".

It has also been pointed out, as an example of the lack of performance of their promise by the nuclear powers that, since the time the NPT was signed, the number of strategic nuclear weapons possessed by the superpowers has increased more than sixfold. Even if the USA and USSR agree to cut them by 50 percent, they would still have more than three times the number they had in 1968.

Moreover, it hardly seems likely that either the nuclear or non-nuclear powers would be prepared to spend the large sums of money necessary to acquire stockpiles of modernized or new nuclear weapons, or start building a stockpile, of untested nuclear weapons. Thus, a CTB could help prevent, or at least curb or inhibit both further vertical and horizontal nuclear proliferation.

Voices have also been raised in recent years maintaining that there was no "bargain" or "quid pro quo" to halt and reverse vertical nuclear proliferation by the nuclear powers in exchange for the non-nuclear states' committing themselves not to engage in horizontal proliferation.

The evidence is clear that there was indeed a mutual "bargain" or "contract" between the nuclear and non-nuclear parties. It has been pointed out on numerous occasions by authoritative government spokesmen that in exchange for the commitment of the non-nuclear parties not to acquire or manufacture nuclear weapons (Article II) and to accept safeguards (Article III), the nuclear parties undertook, in addition to halting and reversing the nuclear arms race (Article VI), also: (a) not to transfer such weapons or control over them to any non-nuclear state or to help it to acquire or manufacture nuclear explosive devices (Article I); (b) to facilitate and promote the peaceful uses of nuclear energy (Article IV); and (c) to assist non-nuclear parties to obtain the peaceful applications of nuclear explosions (Article V).

Article I certainly is not a meaningful or burdensome undertaking since every nuclear power considers it against its interest to permit nuclear weapons to spread to any non-nuclear states and it would oppose this by every means available to it. Article V has never been implemented and most of the nuclear powers appear to be opposed to peaceful nuclear explosions. Article IV has been implemented to some limited extent, but in a manner and at a level that most non-nuclear parties regard as inadequate. Thus, the commitment in Article VI has been traditionally regarded as the most important element of the bargain between the nuclear parties and the non-nuclear parties.

It is also noteworthy that the obligations undertaken in both the PTBT and the NPT were specifically assumed by "each of the parties" to the treaties. Thus, the non-nuclear as well as the nuclear parties are bound to seek a comprehensive test ban and the cessation of the nuclear arms race at an early date.

The question as to whether the NPT in effect embodied a bargain between the nuclear powers and the non-nuclear powers appears to have been effectively settled in the affirmative by many authoritative spokesmen of the nuclear powers themselves, as well as important non-nuclear parties.

The non-nuclear parties feel frustrated and disappointed and, indeed, cheated by the nuclear powers. The non-nuclear parties believe that they have faithfully implemented their commitments whereas the nuclear parties are in flagrant breach of theirs, in particular as regards halting are reversing the nuclear arms race.

Attempts to distort, re-interpret or dilute the clear obligations of the nuclear powers, either as to whether the NPT consitutes a bargain between the nuclear and non-nuclear parties, or as to the linkage between a CTB and the NPT, in my opinion, serve only to anger or irritate the non-nuclear states and are counter-productive.

II. Political Perspective

The basic approach of the non-nuclear states to nuclear proliferation was confirmed at a unique conference of Non-Nuclear-Weapon States that was held in September 1968 following a decision of the General Assembly taken in December 1967. The conference was proposed by Pakistan which was not a member of the ENDC at that time. The work of the conference proceeded in four specific areas:

a) security assurances for the non-nuclear states;
b) measures of nuclear disarmament,
c) prevention of the further proliferation of nuclear weapons by way of safeguards and other measures; and
d) programmes for cooperation in the field of peaceful uses of nuclear energy, including peaceful nuclear explosions.

The four nuclear-weapon powers were invited to participate in the conference but without a right to vote.

Among the various resolutions adopted was one on the cessation of the nuclear arms race and nuclear disarmament. It noted that the achievement of the goal of nuclear non-proliferation requires the adoption of measures to prevent both horizontal and vertical proliferation, and it requested the United Nations to recommend that the ENDC should begin not later than March 1969, negotiations for: (a) the prevention of the further development and improvement of nuclear weapons and their delivery vehicles; (b) the conclusion of a comprehensive test ban treaty; (c) immediate cessation of the production of fissionable materials for weapons purposes and of the manufacture of nuclear weapons; and (d) the reduction and subsequent elimination of all stockpiles of nuclear weapons and their delivery systems.

Although the conference created a great deal of interest at the time, since it was held after the NPT was approved by the General Assembly and opened for signature on 1 July 1968, the conference had little influence on further developments in the field and its work and resolutions were largely ignored and soon forgotten [5].

The few years following the signature of the NPT were mainly concerned with the bilateral SALT negotiations which began in 1969. The Conference of the Committee on Disarmament (CCD), the new name of the committee which had been enlarged in 1969 from 18 to 24 members, was preoccupied with negotiating the 1971 Seabed Treaty and the 1972 Biological Weapons Convention and no multilateral negotiations took place on the question of a CTB. The General Assembly, however, continued each year to call for the suspension of nuclear tests in all environments, and for the CCD to continue as a matter of highest priority its work on a treaty banning underground tests.

After the NPT entered into force in 1970, Article VIII, authorizing a conference of the parties every five years "to review the operation of this treaty with a view to assuring that the purposes of the preamble and the provisions of the treaty are being realized", became more important. The non-nuclear states had agreed in Article III to accept IAEA safeguards to ensure their compliance with the provisions of the Treaty; the review procedure was the only means provided to assure the compliance of the nuclear powers. In practice, the review conferences have served more to assess rather than assure their compliance.

At each of the conferences held in 1975, 1980 and 1985, major efforts were made by the neutral and non-aligned (NNA) parties to persuade the nuclear powers to comply fully with their obligations under the treaty, or at least to improve their performance. At all of them the NNA parties placed main emphasis on Article VI concerning ending and reversing the nuclear arms race. In this regard they called for: (a) an end to underground nuclear testing beginning with a moratorium; and (b) a substantial reduction in nuclear arsenals. Other demands included a pledge not to use or threaten to use nuclear weapons against non-nuclear parties to the treaty and substantial non-discriminatory aid to the developing countries, but it was clear that these were secondary issues. The nuclear parties, for their part, paid most attention to questions of safeguards under Article III and international cooperation in the peaceful uses of nuclear energy.

In each of these review conferences, it was relatively easy to reach a consensus on peaceful uses of nuclear energy and safeguards against diversion, but a consensus on arms control and disarmament was elusive.

In 1975, a fragile over-all consensus was achieved on a Declaration in which the nuclear powers in effect promised to try harder to meet the demands of the non-nuclear states. The Declaration was, however, subject to serious reservations by the NNA which, in an "interpretative statement" attached to the Declaration, said that they stood by their political proposals and interpreted the Declaration in the light of those proposals.

In 1980, however, no consensus was achieved. The nuclear powers would not agree to begin negotiations for a CTB in the Conference on Disarmament, and the conference ended in failure, without any final declaration or even any formal reaffirmation supporting the NPT.

In 1985, because of the escalating nuclear arms race, the UN Secretary-General struck a keynote in his message to the conference in which he stated:

"Unless the nuclear arms race between the major powers is halted and the further spread of military capability deterred, the terrible possibility of wholesale destruction will increase yet further."

In referring to the commitments in Article VI he said,

"In this respect, the implementation of the treaty has been largely one-sided, to the understandable concern and profound dissatisfaction of its non-nuclear-weapon parties. There must be recognition of the fact that restraint on one side cannot reasonably be demanded in the face of unlimited expansion on the other."

As no consensus seemed likely, the NNA members announced they would co-sponsor and press for a vote on three resolutions proposed by Nobel Peace Laureate Ambassador Alfonso Garcia Robles of Mexico. The resolutions called for:

- the resumption in 1985 of negotiations by the three nuclear powers for a comprehensive test ban treaty;
- a moratorium on testing pending the conclusion of a CTB treaty;
- a freeze on the testing, production and deployment of nuclear weapons.

The USA and its allies, fearing that the resolutions would be adopted by the conference, decided to agree to almost any reasonable compromise that would avoid a vote. A compromise was finally worked out that made a consensus possible on a Final Declaration. The compromise was as follows:

"The Conference, except for certain states, deeply regretted that a comprehensive multilateral Nuclear Test Ban Treaty banning all nuclear tests by all states in all environments for all time had not been concluded so far and, therefore, called on the nuclear-weapon states party to the treaty to resume trilateral negotiations in 1985, and called on all the nuclear-weapon states to participate in the urgent negotiation and conclusion of such a treaty as a matter of highest priority in the Conference on Disarmament."

The Declaration also noted that certain states considered deep and verifiable reductions in existing arsenals of nuclear weapons as having the highest priority. (The "certain states" were the USA and UK). It also noted the readiness of the USSR to proceed forthwith to trilateral and multilateral negotiations to conclude a Comprehensive Test Ban Treaty.

It was also agreed that the declaration of the Group of Non-Aligned and Neutral States, as well as the draft resolutions calling for a moratorium on nuclear testing and for a nuclear weapons freeze, should be included as an integral part of the Final Documents.

Although both the 1975 and the 1985 conferences were qualified successes, the latter one differed in an important respect from the former. In 1975, the Non-Aligned and Neutral States could not command a two-thirds majority vote, and they found it necessary to express their reservations in the consensus Final Declaration. In 1985, the Non-Aligned and Neutral States were confident they could obtain a two-thirds majority for their positions and thus were able to have their views reflected in the Final Declaration, while the United States and the United Kingdom found it necessary to express reservations. The 1985 conference demonstrated that voting power had now shifted, or was in the process of doing so, from the nuclear powers and their allies to the NNA states.

As a result of the commitment and persistence of the NNA states, the Final Declaration for the first time focused primary attention on the disarmament aspects of the NPT and stressed the overriding importance of a CTB as having highest priority in the "cessation of the nuclear arms race at an early date".

The non-nuclear states have sent a clear message to the nuclear parties that they must abide by their treaty obligations to halt and reverse the nuclear arms race if the NPT is to endure. It would be in the interest of the nuclear powers as well as of the world if the nuclear powers pay heed.

Because of a fundamental shift in US policy under the Reagan Administration, the USA is now opposed to a CTB, and it will require a major effort to persuade the Bush Administration to revert to the previous US position.

Every US President, from Eisenhower to Carter, favoured a CTB, conditioned only on effective verification, and gave it high priority. Early in the Reagan Administration a CTB was down-graded to a "long-term goal". The US Arms Control and Disarmament Agency stated in September 1983 that nuclear tests were specifically required for the development, modernization and certification of warheads, the maintenance of stockpile reliability and the evaluation of nuclear weapons effects. In September 1985, the Agency stated that before resuming negotiations for a CTB, there must be agreement on "deep reductions in the level of nuclear weapons, maintenance of a credible nuclear deterrent, improved verification capabilities and expanded confidence-building measures". In addition, there have been numerous reports that one of the main reasons for US testing is to develop an X-ray laser for the Strategic Defense Initiative.

In order to play safe, the US also announced that, even if verification of a test ban was effective, it was not in its interest. In 1986 it went to the absurd length of arguing that it had to continue testing so that its allies would retain confidence in the US deterrent; otherwise they might have to build their own nuclear forces to ensure their security.

In September 1987, the USA and the USSR agreed, in a joint statement on Nuclear Testing Talks, to begin long-term step-by-step negotiations not on a test ban but on nuclear testing and verification.

The first step calls for joint experiments that would enable the USA to ratify the 1974 Threshold Test Ban Treaty and the 1976 Peaceful Nuclear Explosions Treaty.

Most countries consider these Treaties as out-dated because of their excessively high thresholds of 150 kilotons and the adequacy of current verification techniques. Moreover, the entry into force of the PNE Treaty would tend to legitimize testing by the nuclear powers, including China and France, and by the non-nuclear states that are not parties to the NPT and who could acquire a nuclear weapons capability under the guise of carrying out explosions for peaceful purposes.

Thereafter, the "ultimate" objective would be a total test ban as part of a process leading to the "ultimate" elimination of nuclear weapons. US officials have freely acknowledged that this programme would not end nuclear testing until some time in the next century at best. They have also admitted that this was a strategy to defuse the growing pressure for a CTB.

In a report to Congress in September 1988 on "The Relationship Between Progress in Other Areas of Arms Control and More Stringent Limitations on Nuclear Testing", President Reagan stated that "nuclear testing is indispensable to maintaining the credible nuclear deterrent which has kept the peace for over 40 years. Thus we do not regard nuclear testing as an evil to be curtailed". The report repeated the previously announced reasons for continuing testing and stated that "Under deep reductions in strategic defensive arms the reliability of our remaining US strategic weapons could be even more important and the need for testing even greater ... "

The report continued

"There is no direct technical linkage between progress in other areas of arms control and the acceptability of progressively more stringent limitations on nuclear testing ... It is clear that limitations as stringent as a complete ban on tests above either 1 kiloton or 10 kilotons yield pose serious risks and will almost certainly not prove to be compatible with our overall security interests ... Thus, even if, in the future, a 1 kiloton threshold were to be effectively verifiable, that fact alone would not make it acceptable or desirable."

It is readily apparent that the latest US position is not only contrary to the traditional US views but that it is in clear conflict with US obligations under both the PTBT and the NPT.

As a result of the inability of the non-nuclear states to persuade the nuclear powers to live up to their obligations under the NPT, there has been a growing disillusionment on the part of the non-nuclear parties, and an erosion of their confidence in both the sincerity of the nuclear powers and in the non-proliferation process itself.

At the same time there is growing concern among the nuclear powers and the industrial states about the proliferation of chemical weapons and sophisticated missiles among developing countries. The latter are becoming wary and suspicious of what they regard as efforts by the nuclear powers and developed countries to extend the non-proliferation process from the nuclear weapons field to these additional areas. Having been "let down" by what they consider a disregard by the nuclear powers of their obligations under the NPT, a number of the developing countries are not willing to put much faith in other non-proliferation regimes. This attitude was evident at the Paris conference in January 1989 of parties to the 1925 Geneva Protocol.

The nuclear powers may now be faced with the task of restoring confidence in their promises and in the concept of non-proliferation. It they persist in their opposition to a CTB, this seems likely to weaken and perhaps undermine the NPT. This, in turn, would tend to erode further the chances of halting the proliferation of other dangerous non-nuclear weapons.

A different reaction seems to be developing among some supporters of a CTB. Because of the adamance of the US opposition to a CTB, some dispirited test ban advocates seem inclined to accept, as a temporary first-step, a low-yield threshold treaty of one, five or ten kilotons, in the hope that these might eventually be reduced to zero. In view of the President's report, however, it is far from clear that such a compromise would be acceptable to the military and the scientists at the weapons-fabricating laboratories. Putting forward such compromise proposals might amount to no more than "negotiating with ourselves" and could have the effect of "undermining the rationale" for a CTB. They would have little or no political impact on the NPT, and the testing countries would maintain the infrastructure for the speedy resumption or increase of testing. Because it is much easier to verify a zero level of testing than any fixed threshold, any threshold treaty would in all likelihood create more verification problems than it would solve and, if the threshold were set at or above 10 kilotons, it could help to promote or preserve the continued maintenance of concepts of nuclear superiority and war-fighting strategies. Instead of representing a step towards a CTB, it could have the opposite effect of preventing or delaying a CTB. Judging by past experience with other partial treaties, a low-yield threshold treaty would tend to erode the pressures from the public for a CTB within the USA and other nuclear powers, and could encourage the Chinese and the French to continue their testing.

The history of the nuclear arms control negotiations shows that no first-step or partial arms limitation or reduction treaty has been followed by further ratified agreements to limit, reduce or eliminate the weapons concerned.

Thus, the effect of a first-step or interim threshold agreement might well be that negotiations for a CTB would be abandoned, not just for several years, but for several decades. There would be no assurance whatsoever that the CTB negotiations would be resumed in the foreseeable future or that any further interim agreements could be achieved scaling down either the number or yield of future tests. It would seem to me that it would be far better to hold out for and maintain the pressure for a CTB [6].

III. Conference to Amend the PTBT

Disillusioned and discontented by the failure of the nuclear powers to live up to their promises, a movement was started three years ago among the 113 non-nuclear parties to the PTBT to call for a conference to amend the Treaty and ban underground tests. The Treaty provides that any party may propose an amendment; if one-third of the parties support the request, the three "depositary" governments (the United States, Great Britain and the Soviet Union) "shall" convene a conference to consider it.

Shocked by the latest actions of the superpowers, a group of Non-Aligned countries, led by Mexico, concluded that the time had come to act. They proposed that the UN General Assembly support a proposal to amend the 1963 Treaty in order to convert it into a CTB treaty.

In November 1987, the idea was endorsed by the General Assembly, which recommended that the non-nuclear signatories formally submit a proposal to begin the amendment process. In all, 128 countries including the Soviet Union voted for the recommendation. Only the United States, the United Kingdom and France (which is not a party to the Treaty) voted against it. On August 5, 1988 – the 25th anniversary of the signing of the Partial Test Ban Treaty – six parties to the Treaty – Indonesia, Mexico, Peru, Sri Lanka, Venezuela and Yugoslavia – formally proposed that a conference be convened to convert the PTBT into a CTB. The move was enthusiastically welcomed by the United Nations General Assembly.

At long last it now appears certain that multilateral negotiations to stop *all* nuclear weapon testing will take place. For the first time in the nuclear age, a number of non-nuclear weapon states have taken matters into their own hands in an effort to stop underground testing and thus curb both vertical and horizontal nuclear weapon proliferation.

The amendment provision was inserted in the PTBT, among other reasons to attract the adherence of the largest number of non-nuclear states. The Treaty now has 116 parties and, since more than the required 39 requests have been made, the convening of the conference is mandatory. The three nuclear powers have announced that they will abide by their Treaty obligations.

The sponsors of the amendment want the conference to continue to meet until agreement is reached with the nuclear powers on a verifiable ban on underground tests. The task will not be accomplished quickly or easily. At the conference, the United States will in all likelihood, at least in the beginning, oppose the amendment, and it may take several years before it is persuaded. This may have to await the outcome of the 1990 mid-term congressional elections, which usually go against the Administration in office. Although only a majority of the parties (58) must approve an amendment in order for it to enter into force, the three nuclear powers must be included in that majority, and they thus have a veto. While the Soviet Union supports the amendment, the United States and the United Kingdom are at present opposed and for years have blocked all attempts to begin multilateral negotiations. There are, however, some hopes that the new Bush Administration may adopt a different attitude.

Some of these hopes are based on the fact that the USA, which has conducted many more tests than the USSR, and clearly has acquired greater technical sophistication, could lock in its advantage if it would agree to a CTB. In fact, many ardent American advocates of the CTB are puzzled as to why the USA has not adopted that course in the past and why, if in fact the arms race is abating, it does not do so now.

Reason and logic are on the side of ending all tests. Both superpowers now realize that it is in their own interest to cut back the arms race and divert their astronomically high military expenditures to human needs. Moreover, contrary to the views expressed in President Reagan's report, continuing improvement in American-Soviet relations should create a positive mutual feed-back between a total test ban and deep cuts in strategic weapons.

The non-nuclear states are confident that their amendment initiative will eventually succeed. The very fact that negotiations are proceeding at the conference to end all tests will generate strong national and international pressures on the United States and Great Britain, where public opinion polls strongly favour a total ban.

Most important, the link between a total test ban and the NPT provides the strongest argument in support of the test ban.

The NPT comes up for its final five-year review in 1990. As mentioned, in all three previous review conferences, the non-nuclear states gave top priority to a comprehensive test ban. Most non-nuclear states consider a total test ban as the minimum price that the nuclear powers must pay to ensure the continued viability of the NPT. If the USA and UK reject an amendment for a complete ban, then the NPT, one of the most important and successful treaties of the nuclear age, might itself be in serious danger when it comes up for renewal in 1995. Proposals are already being floated to negotiate a new NPT or amend the present one, and to convene another conference of non-nuclear weapon states. These prospects give the nuclear powers nightmares.

A comprehensive test ban would be the surest way to strengthen the NPT and the entire non-proliferation regime. The PTBT contains a unique provision that is not present in any other arms control treaty: if an amendment is approved by a majority of the parties, including the three nuclear powers, it automatically becomes binding on all of the parties. It so happens that all the near-nuclear states are parties to the Partial Test Ban Treaty, including Argentina, Brazil, India, Israel, Pakistan and South Africa, although none of these six is a party to the NPT [7].

Hence, if the amendment is ratified by 55 parties plus the three nuclear powers, all of these near-nuclear states would be bound and would not be able to conduct any nuclear tests in order to "go nuclear". Thus, a total test ban would be the easiest way to reinforce the non-proliferation regime and should provide a powerful inducement to the nuclear powers to ratify the amendment.

Because of the combination of disincentives to block the proposed amendment and of incentives to support it, the amendment conference could achieve a breakthrough in the efforts to control the nuclear arms race. The mere holding of the conference would, for the first time, provide an opportunity for all concerned countries to participate in negotiations for an end to testing. It would be a step to reduce the exclusive monopoly of the nuclear powers over nuclear arms control and make them more accountable to the community of nations. It would also demonstrate the revival of multilateralism as a part of the current revitalization of the United Nations.

Supporters of the NPT quite rightly keep emphasizing the importance of strengthening the treaty. But their proposals for doing so are not very persuasive. They consist in the main of urging that the countries that have not joined or acceded to the NPT should do so. While this is a commendable and desirable aim, its chances of persuading the "holdouts", the six near-nuclear countries mentioned above, to do so are minimal. These states believe that they have strong reasons for remaining outside of the Treaty. And as the Treaty approaches the end of its initial period in 1995, they tend to regard their attitudes as having been confirmed and not least by the failure of the nuclear powers to carry out their obligations.

The NPT has, without doubt, been one of the most important disarmament treaties in the sense that it has been the main pillar of the non-proliferation regime and in preventing the horizontal proliferation of nuclear weapons. It already has more parties than any other arms control treaty, and the number of parties keeps increasing. But almost all of the new parties are smaller developing countries. Thus, the efforts of active supporters of the NPT who down-play the importance of the CTB to get more countries to accede to the Treaty may be self-defeating; since the Non-Aligned Movement has twice endorsed the amendment approach, these new parties may well join the majority in insisting at the 1990 review conference and the 1995 renewal conference that the CTB is the best way of strengthening the NPT.

It is possible that the nuclear powers and their allies who are strong supporters of the NPT, but who oppose a CTB, are deluding themselves if they think that the NPT can be strengthened or even "saved" without a CTB. Some of them argue that it is better to save the NPT even if a CTB must be sacrificed. On the other hand, a number of non-nuclear states that support both the CTB and the NPT maintain that the most likely prospect is that we shall have both a CTB and an NPT or neither.

There is no guarantee in view of the prevailing mood, that even a CTB, by itself, would be sufficient to ensure the continued viability of the NPT, but I fear that, without at least the achievement of a CTB, the NPT will be in serious jeopardy. I believe that it would be the course of wisdom to do everything possible to help achieve a CTB both for its own sake and as the best, and perhaps the only, way to really strengthen the NPT.

In any case, the USA and UK must balance the risk of rejecting a comprehensive test ban at the amendment conference, with its negative impact on the

NPT, against what the overwhelming majority of states believe are the lesser risks of giving up testing and the unending, futile and increasingly dangerous nuclear arms race.

References

1 G. T. Seaborg, *Kennedy, Khrushchev and the Test Ban* (University of California Press, 1981) p. 249

2 In a national radio-television address by President John F. Kennedy on July 26, 1963, after the initialing of the agreed text of the PTBT the previous day, he stated: "This can be a step toward preventing the spread of nuclear weapons to nations not now possessing them ... This treaty can be the opening wedge in that campaign ... " (US ACDA, Documents on Disarmament, 1963, pp. 254–255)

3 On the negotiations related to Article VI, see W. Epstein, *The Last Chance: Nuclear Proliferation and Arms Control* (New York: Macmillan Free Press, 1976) pp. 75–79, 95–95, and 181–200

4 US Ambassador Arthur Goldberg, former Justice of the Supreme Court of the United States, in his statement to the First Committee on April 26, 1968, during the debate on the NPT, stated in regard to Article VI: "The language of this article indicates a practical order of priorities ... headed by 'cessation of the nuclear arms race at an early date' and proceeding next to 'nuclear disarmament' and finally to 'general and complete disarmament ... ' " (US ACDA, Documents on Disarmament, 1968, p. 230)

5 See "The Forgotten Conference of the Non-Nuclear States", Epstein, op. cit., Chapter 9, pp. 126–134; "The Conference of Non-Nuclear-Weapon States", *The United Nations and Disarmament 1945–1970* (United Nations, 1970), Chapter 14, pp. 307–326

6 For further elaboration of this point of view, see Steve Fetter, *Toward a Comprehensive Test Ban* (Cambridge, MA: Ballinger Publishing Co., 1988)

7 In this connection, the advantages of using the amendment procedure of the PTBT in order to strengthen the NPT were recognized by the US Arms Control and Disarmament Agency. In 1977, it considered a suggestion by the Assistant Deputy Director in charge of Non-Proliferation to use the amendment procedure as a means of achieving a comprehensive test ban and thus strengthening the NPT. The idea was not pursued, however, because Argentina and Pakistan had not yet become parties to the PTBT, and it was feared that Brazil and India, who were parties, and thus would be bound by a comprehensive test ban amendment, might regard the amendment approach as discriminatory and unfriendly

A Comprehensive Test Ban and the Prevention of Horizontal Nuclear Proliferation

Vitalii I. Goldanskii and Valerii F. Davydov

To appreciate the importance of a Comprehensive Test Ban Treaty (CTB) for the prevention of horizontal nuclear weapons proliferation it is necessary first of all to outline the present situation in this field of international security.

By the end of the 1980s one of the alarming trends in international relations was the mounting risk of an enlargement of the "nuclear club". At present, when the United States and the Soviet Union – both depositaries of the Non-Proliferation Treaty – are realizing the necessity of neutralizing the threat from their mutually opposing nuclear arsenals and are committed to their drastic reduction, or even total abolition, the nuclear arms race threatens to accelerate and follow a path already trodden by the five nuclear powers.

At present there are more than 140 states party to the 1968 Treaty on the Non-Proliferation of Nuclear Weapons. Among them are three nuclear states – USA, USSR and the United Kingdom – and also a large group of countries with highly developed nuclear industries capable of producing nuclear weapons at short notice. These "threshold" states undertook, according to the Treaty, a solemn commitment not to follow the nuclear weapons path. There are about 40 states outside the Non-Proliferation Treaty, including two nuclear-weapon states – France and the People's Republic of China – and the threshold states of Israel, India, Pakistan, South Africa, Argentina and Brazil. These latter states have the technological potential to produce nuclear weapons and may do so without violating any political obligation.

All the main candidates to the "nuclear club" – Israel, Pakistan and South Africa – are undertaking intensive preparatory steps in the nuclear field. Some of these states are classified by experts as "clandestine" nuclear states that have already secretly created nuclear devices. Other analyses differ only as to how many weapons they can produce and when. All of the threshold states closely monitor the problem of accumulating fissionable materials and the acquisition of nuclear technology as well as the development or acquisition of effective nuclear-weapon delivery vehicles. These states maintain the option of conducting nuclear tests as a means of openly declaring their nuclear status.

In view of the seriousness of horizontal proliferation, the international political agenda now has to include the question of nuclear conflict between threshold states, measures to curb nuclear terrorism, and to safeguard nuclear facilities from military action, as well as means of limiting international trade in missile technology. It is understandable why most experts think that horizontal proliferation is as serious a threat to international security as vertical proliferation. Some experts even consider that if nuclear weapons are used again they will be employed in conflicts between future nuclear states rather than in a Soviet-American confrontation.

The emergence of horizontal proliferation into an imminent and real danger stems to a large extent from the cult of nuclear weapons engendered by the nuclear powers during the post-war period.

The creation of huge nuclear arsenals capable of annihilating the world several times over, the readiness to use them, and the continuation of nuclear testing testified, and continues to testify, that nuclear weapons represent a non-negotiable military and political value for the support of national interests. This same attitude also underlies the thinking of the threshold states regarding nuclear weapons. Given this reality, what can be done to stop or substantially slow down horizontal proliferation? The technical hurdles to nuclear weapon development have been largely overcome by the threshold states, and existing political limitations have not been effective. The only available path is to start to modify the international environment surrounding the threshold states and to rid it of the nuclear weapon factor. This means placing a reliable barrier in the way of vertical proliferation of nuclear weapons. And this political barrier can only be a Comprehensive Test Ban Treaty.

The NPT and Nuclear Testing

The Treaty on the Non-Proliferation of Nuclear Weapons continues to be the cornerstone of international policy on nuclear proliferation. Despite the ever-growing number of states party, its fate is not yet certain. The Treaty expires in 1995 when the states party will have to decide whether to prolong it or not. Most experts are not fully confident about a satisfactory solution to the problems facing the Treaty. In particular, there is a substantial difference of opinion between the three nuclear states party, on the one side, and the majority of non-nuclear states party, on the other, over the estimate of its value and effectiveness. This disagreement stems first of all from the problem of the mutual fulfillment of their obligations under the Treaty.

In the sixties, when the Treaty was being negotiated, there was a consensus on the necessity to balance the obligations of the non-nuclear and nuclear states party to the Treaty. Thus Article II requires the former "not to manufacture or otherwise acquire nuclear weapons or other nuclear explosive devices" while the latter are obliged by Article VI "to pursue negotiations in good faith on effective measures relating to cessation of the nuclear arms race at an early date and to nuclear disarmament". Moreover, the Preamble to the Treaty states the main way to curb the nuclear arms race by calling on the nuclear powers to proclaim their intention "to achieve the discontinuance of all test explosions of nuclear weapons for all time" [1].

Thus, the direct linkage between horizontal and vertical proliferation was made when it was evident that success in attaining the first goal depends on the successful outcome of the second. All of the nuclear powers officially recognized this linkage. Ambassador Arthur Goldberg of the United States, in his statement before the United Nations in 1968, said: "The permanent viability of this treaty will depend in large measure on our success in further negotiations contemplated in Article VI" [2].

The nuclear-weapon state depositary governments of the USA, USSR, and the United Kingdom also understood that a CTB would be the main evidence of their readiness to fulfill their obligations. President Carter underlined this: "A comprehensive test ban would also signal to the world the determination of the signatory states to call a halt to the further development of nuclear weaponry" [3].

Yet, official recognition of the linkage between horizontal and vertical proliferation was not reflected in the policy of the nuclear-weapon state depositary governments until recently. Twenty years after the signing of the NPT it is not difficult to conclude that, contrary to the non-nuclear states, which have been and are abiding by their obligations, the nuclear-weapon states party to the Treaty were for a long time ignoring theirs. From 1968 to 1988 their nuclear arsenals multiplied manyfold so that they now contain 50,000 nuclear explosive devices. At the same time not only the quantitative but also the qualitative arms race was going on. After the 1963 Partial Test Ban Treaty, their yearly testing rate rose dramatically. The absence of a CTB opened the way to an ever-spiralling arms race for the nuclear powers, and made Soviet-American arms control and limitation talks in the seventies a Sisyphus task.

The nuclear powers' approach to the problems of disarmament often attracted the criticism of the non-nuclear states. During the NPT Review Conferences of 1975, 1980 and 1985, non-nuclear states were constantly stressing the importance of a CTB as one of the first steps leading to a strengthening of the global nuclear non-proliferation regime. At the first NPT Review Conference in 1975 great efforts were made to reach a consensus on support of the non-proliferation regime. The second conference in 1980 ended in complete failure despite the fact that the USA, USSR and the United Kingdom were very close to signing a CTB at the time. It even failed to adopt a final declaration on NPT support. The 1985 conference faced a similar situation. The anxiety of the majority of the non-nuclear states party to the NPT over the rising momentum of the arms race during the period of 1980–85 brought the conference to a stalemate. Only deft diplomatic manoeuvres and the cooperation between the Soviet Union and the United States saved it from ending in failure.

The United States and the United Kingdom found themselves in complete isolation not only from the non-aligned and neutral states but even from the majority of their own allies over the issue of a comprehensive test ban. Their declarations in favour of deep reductions in nuclear weapons as a means of fulfilling their obligations under Article VI made no impression on the majority of conference participants after their refusal to join the Soviet Union's 1985 moratorium on nuclear testing and to begin negotiations on a CTB called for by the USSR.

Despite American pressure, a group of non-aligned and neutral states prepared a declaration which "called" upon the three depositary states of the NPT to institute, as a provisional measure, an immediate "moratorium on all nuclear-weapon tests" and also "to agree on a complete freeze on the testing, production and deployment of all nuclear weapons and delivery vehicles".

This declaration was incorporated into the Final Declaration as a complementary but nonetheless integral part. At the insistence of the majority of the conference participants, the Final Declaration contained appeals – toned down to be acceptable to the USA and the UK – which more realistically reflected the state of affairs surrounding Article VI:

"The Conference reiterated its conviction that the objectives of Article VI remained unfulfilled and concluded that the nuclear-weapon States should make greater efforts to ensure effective measures for the cessation of the nuclear arms race at an early date, for nuclear disarmament and for a Treaty on general and complete disarmament under strict and effective international control ...

The Conference, except for certain States ... deeply regretted that a comprehensive multilateral Nuclear Test Ban Treaty banning all nuclear tests by all States in all environments for all time had not been concluded so far and, therefore, called on the nuclear-weapon States Party to the Treaty to resume trilateral negotiations in 1985, and called on all the nuclear-weapon States to participate in the urgent negotiation and conclusion of such a Treaty as a matter of the highest priority in the Conference on Disarmament" [4].

In the eyes of the non-nuclear states party to the NPT it is a CTB that was and continues to be proof of the nuclear states' willingness to fulfill their obligations in the field of disarmament. Certainly the 1987 Soviet-American INF Treaty and their current proposals for a fifty percent reduction in their strategic nuclear weapons will have full backing of the non-nuclear states party to the Treaty as a first step towards the liquidation of nuclear arsenals. But at the same time, in the absence of a CTB, they may legitimately ask whether the nuclear powers are really trying to stop the nuclear arms race by making some quantitative reductions. The fact is the USA refuses to accept an unconditional nuclear test ban and is preparing to deploy third-generation nuclear weapons. And if this is so, this is evidence of a continuing disregard of nuclear disarmament obligations.

How will this affect the Treaty's fate? The indefinite continuation of the Treaty is often taken for granted. To some extent this point of view reflects the conclusion of the majority of states party to the Treaty that nuclear weapons weaken their national security rather than strengthen it. But this does not exclude the possibility that certain countries will conclude that their national interests will be better served by the acquisition of nuclear weapons and will withdraw from the Treaty. Article X provides for this contingency. If some states have reservations on this issue, it would be more beneficial to them to use their temporary participation in the Treaty as a cover for the acquisition of nuclear technology and material stockpiling, before openly acquiring nuclear weapons under the pretext that the nuclear states party to the Treaty are not abiding by their obligations under it.

Another more important question is how long will the non-nuclear states tolerate the nuclear states' neglect of their obligations under the Treaty? It would be too extreme to conclude that the non-proliferation regime will cease to exist in 1995 when the Treaty expires. The experience of the review conferences indicates that the nuclear powers and their allies are capable of overcoming their differences and are able to cooperate in preserving the non-proliferation regime. But it would be a mistake to assume that the Treaty can be kept intact as it is without clearer obligations regarding a nuclear test ban. The fact is the non-nuclear states party can put amendments into the Treaty's text, including those dealing with a comprehensive test ban, regardless of the nuclear powers. If such changes are rejected by the nuclear powers then the NPT will proably be fatally compromised with all the negative consequences this would entail for international security.

The Nuclear-Weapon States and Nuclear Testing

The Soviet-American talks that began in 1987 on a limited nuclear test ban with a view towards a comprehensive one, give hope that the crucial problem on which the fate of the NPT depends can be satisfactorily solved. But this hope can be justified only if the United States and the Soviet Union jointly proclaim the urgency of a comprehensive test ban and undertake real steps to attain it. This cannot be counted on at present. The Reagan Administration, unlike the Soviet Union which is ready for an immediate test ban, continued to stress the importance of nuclear tests and only recently agreed to consider their ban as a long-term goal. In a 1987 statement to Congress, President Reagan said: "A Comprehensive test ban remains a long-term objective of the United States. Such a ban must be viewed, however, in the contexts of a time when we do not need to depend on nuclear deterrence to ensure international security and stability, and when we have achieved broad, deep and verifiable arms reduction, confidence-building measures, and achieved greater balance in conventional forces" [5]. Such heavily qualified statements do not lend any hope for an agreement on a comprehensive test ban even in the distant future.

The Reagan Administration maintained that it was necessary to proceed stepwise in the test-ban process, beginning with the ratification of the Soviet-American Treaties of 1974 and 1976 on a threshold underground test ban. What role can the ratification of the Soviet-American Treaty on the Limitation of Underground Nuclear Weapon Tests of 1974 and the Treaty on Underground Nuclear Explosions for Peaceful Purposes of 1976 play in the prevention of horizontal proliferation? From the point of view of non-nuclear countries, the 1974 Treaty indicates an intention to somehow regulate the nuclear arms race by emphasizing further sophistication and miniaturization of nuclear warheads. The further reduction of the threshold for nuclear tests and the application of quotas on these tests may be viewed as a desire to safeguard the effectiveness of nuclear arsenals and to enhance the performance of tactical nuclear weapons. It is clear that the conclusion of agreements on threshold nuclear test bans between the USA and USSR can hardly be compared to one on a comprehensive test ban. It is not surprising that the majority of the non-nuclear states are indifferent at best toward the fate of the 1974 Treaty. At worst they suspect that the Treaty's framework is flexible enough to sustain research on tactical nuclear weapons which they see as the main threat to their security. They would welcome its ratification only if they believed it would be followed by a CTB. The non-nuclear states' suspicion of a threshold test-ban treaty is not without basis. It is plausible that after the elimination of intermediate and shorter-range missiles and a fifty percent reduction in strategic nuclear forces, the tactical nuclear arsenals on both sides will grow in the absence of a Soviet-American agreement on their limits.

The non-nuclear states' attitude towards the 1976 Treaty on Underground Nuclear Explosion for Peaceful Purposes is a contentious one. Some, especially those not party to the NPT, continue to stress their right to conduct such explosions. Bearing in mind that there is no clear distinction between a peaceful and a military test explosion, the legalization of peaceful explosions by the ratification of the 1976 Treaty will be detrimental to the non-proliferation cause. Under the pretext of conducting peaceful nuclear explosions, threshold states can openly enter the nuclear arms race. Consequently, a total moratorium by the USA and USSR on all peaceful underground nuclear explosions would be a meaningful step. This approach is supported by a large group of states party to the NPT, including the industrialized countries. As in the case of the 1974 Treaty, the ratification of the 1976 Treaty only makes sense if it is an intermediate step towards a ban on peaceful nuclear explosions.

Certainly, the step-by-step approach to the realization of a test ban is preferable to just marking time. Taking into account the interests of the non-proliferation regime and the increasing threat of nuclear tests by the threshold states, a CTB should have priority and be the focal point of Soviet-American talks in Geneva on nuclear testing.

Will the three depositary states party to the NPT have time to come to an agreement on a comprehensive test ban before the Treaty expires? If not, is a NPT crisis inevitable? This is of crucial importance to the future of the non-proliferation regime. And another question: Will one of the threshold states openly acquire nuclear weapons before the nuclear powers have finished their negotiations?

A CTB's Impact on Horizontal Proliferation

The opponents of a CTB often state that the conclusion of such a treaty would aid the horizontal proliferation of nuclear weapons. In the wake of a ban on the further sophistication of American nuclear arsenals and a weakening of nuclear guarantees, America's allies might produce their own nuclear weapons to replace the American nuclear umbrella. Although such an argument might serve as an example of formal logic, it is too contradictory as far as the real aspirations of the majority of America's allies are concerned. All of America's allies in Western Europe, Japan and the ANZUS member states support Soviet-American zero-level INF and SNF agreements and fifty-percent reductions in strategic nuclear offensive forces; they have not expressed any reservations about the reliability of American deterrence vis-à-vis a CTB. On the contrary, America's allies are far more concerned about the nuclear arms race getting out of control and leading to a nuclear catastrophe in which they would be the first and main victims. The policy of the majority of America's allies to distance themselves from American nuclear planning and to forbid the deployment of nuclear weapons on their soil received wide popular support. These allies included Denmark, Norway, Iceland, Spain, Greece, Portugal, Australia, New Zealand and the Philippines. Japan continues to abide by the principle "not to introduce, not to produce and not to acquire" nuclear weapons. West Germany's main opposition party, the Social Democrats, campaigns against the deployment of American nuclear weapons in the FRG. American allies as well as Soviet ones support the non-proliferation regime and are parties to the Treaty.

Moreover, the overwhelming majority of these allies as well as communist, non-aligned and neutral nations, support a comprehensive test ban as a means of halting horizontal and vertical proliferation. The head of the Japanese delegation, R. Imai, speaking at the Third NPT Review Conference, expressed this view in no uncertain terms: "My country has stressed the importance of the test ban as an important first step on the road to nuclear disarmament and thus objects to any nuclear tests by any state. A test ban is of importance in preventing the vertical as well as horizontal proliferation of nuclear weapons and would thus be an important complement of the NPT regime" [6].

The non-nuclear states' emerging consensus on a comprehensive nuclear test ban, despite their alliance memberships, is becoming a reality in international relations. Proof of this is reflected in the unanimous support by the overwhelming majority of the United Nations' member states of resolutions calling for a comprehensive nuclear test ban and for the signing of a CTB.

In the political sphere, the signing of a CTB between the USA, USSR and UK would lead to a strengthening of their positions as states-founders of the non-proliferation regime and would facilitate universal adherence to the NPT. France and the People's Republic of China would hardly be able to indefinitely ignore the CTB's existence. It is worth noting that after the signing of the 1963 Partial Test Ban Treaty, France has not conducted any atmospheric tests since 1975, nor has China since 1981.

Since the mid-eighties, the French nuclear tests on Mururoa Atoll have aroused an ever-growing wave of protests from the states of Asia and the Pacific region which are resolved to force Paris to end such tests. Protocol 3 of the Rarotonga Treaty establishing a nuclear-weapon-free zone in the South Pacific forbids the testing of any nuclear explosive device within the zone. Although France, as well as the USA and UK, refuse to join the Treaty, nevertheless they have to come to terms with the state parties to the nuclear-free zone on this subject. Describing France's approach to nuclear disarmament, French officials under the pressure of international public opinion were forced to state that Paris would join talks on the reduction of its nuclear arsenal after the Soviet Union and the United States radically reduced theirs. In brief, France's ability to sustain its nuclear testing programme, even for a short time, in the face of a comprehensive test ban appears to be limited. It is well to recall here that in spite of its refusal to join the NPT, France has officially stated that it would act as if it were a state party.

Experts agree that the People's Republic of China's nuclear force modernization programme has been comparatively moderate. From 1964 to 1987 the PRC conducted a total of 30 nuclear explosions. In 1987 China acceded to Protocol 3 of the Rarotonga Treaty. Having declared that it will not be the first to use nuclear weapons and that it will not use them against non-nuclear states, the PRC repeatedly expressed its official support of American and Soviet nuclear force reductions during the 1980s. Like France, the PRC declared that it would act as though it were a party to the NPT. Taking into account China's long-standing political links with the developing and neutral states, which support a test ban, it is evident that China's interests would not be served by ignoring a test ban treaty. It would rather join it than have its long-term interests jeopardized in the developing world.

The Threshold States' Response to a CTB

The importance of a CTB for the prevention of horizontal proliferation becomes particularly transparent if we undertake a case-by-case study of the threshold states which remain outside the NPT.

By the end of the 1980s, India will be facing an ever-growing challenge to make a choice: whether to openly acquire nuclear weapons or to become a leader of the neutral and non-aligned states' movement toward nuclear disarmament, while maintaining the capability to acquire nuclear weapons should its policy take a different turn. It is often stated that whether India becomes a nuclear power or not depends mainly on the conduct of China and its main rival, Pakistan. But it seems that this is just one of many factors influencing India's position. The decisive one is the Soviet-American arms race. India's final choice will depend on whether it is halted and reversed, including the elimination of nuclear weapons as a factor in international relations.

India's refusal to join the NPT has always been predicated on the supposition that it finalizes the division of the world's states into nuclear and non-nuclear ones, that it sanctions the nuclear arms race and excludes non-nuclear states from it. This discrimination is symbolized by the continuation of nuclear testing and the advancement of nuclear states by further nuclear weapon modernization and sophistication. To a large extent this factor explains India's negative attitude to Pakistan's proposal of joint entry into the NPT, a bilateral agreement not to develop nuclear weapons, mutual inspection of nuclear facilities, and the creation of a nuclear-free zone in South Asia. At the same time, India has constantly supported calls for the curbing of the global arms race, including Soviet-American agreement on INF and SNF, fifty-percent reductions in strategic offensive weapons followed by the total elimination of nuclear arsenals, and international agreements on the non-use of nuclear weapons. Among the many proposals in this area, a CTB has received especially strong support.

Since the mid-fifties India has strongly supported such a treaty. Delhi has regularly supported initiatives to eliminate nuclear weapons and voted for General Assembly resolutions to stop the arms race and to ban nuclear testing. India joined the 1963 Partial Test Ban Treaty. When the Soviet Union declared a moratorium on nuclear testing in 1985, India headed the Six Nation Initiative that called on the other nuclear powers to follow the Soviet example. Moreover, India, along with its partners, offered to help verify a CTB.

After conducting its 1974 "demonstrative" nuclear explosion, India froze its nuclear preparations. The Ghandi government's position of "restraint" in the nuclear field is under strong pressure from those political and military circles which believe that India only loses time by "asking" nuclear powers to start radical reductions of nuclear weapons, that India's status in world politics and its security vis-à-vis the USA, Pakistan and China would only be enhanced if it openly set out to acquire nuclear weapons and began a nuclear testing programme. The director of New Dehli's Institute of Defence Studies and Analysis, K. Subrahmanyam, underlined the utility of this policy: "... India should exercise its nuclear options before 1995, before the so-called Non-Proliferation Treaty is attempted to be made permanent" [7].

In the absence of a CTB the influence of the supporters of India's possession of nuclear weapons on official government policy grows. The majority of experts believe that India is continuing to accumulate fissionable materials and to acquire technology that can be used for military purposes. Great attention is paid to the development of missile technology that can also be applied to nuclear-weapon delivery vehicles. The present unstable balance between nuclear and non-nuclear options can be not only maintained but tipped in favour of non-possession of nuclear weapons by the conclusion of a universal CTB.

Experts agree that Pakistan will have accumulated the materials and mastered the technology necessary for the production of nuclear weapons by the end of the 1980s. Pakistan's late President Zia ul-Haq in an interview with *Time Magazine* published on March 30, 1987 openly stated: "You can virtually write today that Pakistan can build a (nuclear) bomb whenever it wishes" [8]. With the help of covert operations, Pakistan bought materials in Western markets needed for the development of non-nuclear components of nuclear explosive devices. Pakistan's armed forces have nuclear-capable F-16s and Mirage aircraft, and they are active in the field of missile technology.

The Reagan Administration rejected the Symington amendment in 1981 on the pretext that the situation in Afghanistan required that Pakistan be given massive military aid and political support, eliminating a major obstacle to its nuclear ambitions. During the 1980s a cat-and-mouse game of Washington asking Islamabad to halt nuclear weapon development lest it compromise itself as a champion of non-proliferation, and of Islamabad denying such intentions while secretly pursuing a nuclear policy, played itself out with Pakistan the victor. Its leaders skillfully manipulated America's interest in strengthening its politico-military ties with Pakistan in order to establish a bridgehead for action against Afghanistan.

Consequently, the United States seems to have lost its leverage to influence the future development of Pakistan's nuclear policy. Not even a Congressional moratorium on further military aid to Pakistan is likely to enlighten the Pakistani ruling elite and result in some visible change in their conduct on nuclear matters.

Pakistani leaders have repeatedly stated that if some neighbouring country – India – conducts another nuclear test, the Pakistani people "would eat grass" in order to acquire nuclear weapons. In the 1980s Western media repeatedly reported that Islamabad was preparing for nuclear tests. Are such tests possible in the near future? If India should test again then Pakistan will immediately respond. But even if India shows restraint it is hard to give a definite "yes" or "no" answer. On the one hand, Pakistan would hardly want to provoke India, thereby starting an open nuclear arms race in South Asia. Pakistani leaders know that Congress would halt American military aid. And Soviet-Pakistani relations would be greatly strained. There may also be technical impediments to testing at this stage of its nuclear programme. On the other hand, Pakistan's ambitious efforts to become a leader of the Islamic world and to develop not only a Pakistani but also an "Islamic bomb", put strong pressure on the nuclear option, despite pragmatic considerations and all the negative consequences this step would entail for Pakistan. From this viewpoint, the problem is not whether but when to conduct nuclear tests and under what pretext.

As in India's case, a global test ban would be the most efficient, perhaps the sole, means of preventing Pakistan from acquiring nuclear weapons. Although Pakistan has not joined the Partial Test Ban Treaty, it regularly votes in the United Nations for resolutions calling for a CTB and officially supports the non-use of nuclear weapons and their elimination. Nuclear rivalry in South Asia inevitably makes this region an international flash point where there is an increasing risk of conflict between "non-nuclear" powers and subsequent superpower involvement. A CTB would be a modest price to pay for ending this menace.

Israel presents a clear case of a state striving to acquire fissionable materials and nuclear technology. Most experts agree that Israel became a clandestine nuclear state long ago. Evidence presented by the Israeli nuclear technician

M. Vanunu in 1986 indicates that Israel might have 100–200 nuclear explosive devices at its disposal. Moreover, some experts believe that Israel has already developed and produced a thermonuclear bomb without having to conduct any tests. In 1987 its Jericho missile was tested in the Mediterranian over a range that brought it close to the 1000 km intermediate range class.

Israel's refusal to join the NPT is motivated by its contention that the Treaty cannot effectively prevent a nuclear weapon breakout by a state party. It has repeatedly stated that a whole range of Arab countries, including Libya and Iraq, joined the NPT with the aim of acquiring a nuclear potential that can be switched to military purposes whenever they wish. Israel's bombing of Iraq's peaceful nuclear facility in 1981 showed that Tel-Aviv is ready to resort to extreme measures, including the use of nuclear weapons, if another neighbouring country creates them.

At the same time, Israeli ruling circles admit to some extent that Arab countries will not indefinitely allow Israel to monopolize its status as the only clandestine nuclear state in the region. The majority of Arab countries, and also Iran, show an increasing interest in acquiring nuclear technology. One may cite the frequent treaties of cooperation in the nuclear field concluded between Arab states and Argentina and Brazil. Although Arab countries are not likely to match Israel's nuclear programme in the near future, their aim is clear.

In order to head off future reprisals from Arab states Israeli rulers are carefully trying to avoid attracting international attention to the growth of Israel's nuclear potential. It officially stresses that it will neither be the "first to introduce nuclear weapons" in the region nor again destroy the nuclear facilities of neighbouring countries, and that it supports the establishment of a nuclear-free zone in the Middle East. Fearing a nuclear weapon breakout in the Arab countries and also in Iran, Pakistan and India, Israel has expressed some interest in supplementary measures that would freeze the *status quo* in the nuclear field. A CTB might be such a measure.

Israel is a party to the 1963 Partial Test Ban Treaty. Further sophistication of Israel's nuclear potential may bring nuclear testing to the fore. Israel may risk it, especially if a threshold state conducts a nuclear test to declare itself a nuclear power. So far the negative political consequences of a nuclear test have prevented Israel from conducting one.

The majority of experts agree that South Africa, like Israel, can also be called a clandestine nuclear state. Discrepancies arise only as to how many bombs Pretoria is capable of manufacturing from the fissionable materials it has accumulated. A figure of 11 – 18 is sometimes mentioned, and sometimes a larger one [9]. Pretoria has never tried to conceal the fact that it wants to acquire nuclear weapons. After India's 1974 test, South Africa was very close to testing in 1977. The facilities being constructed in the Kalahari Desert detected by space satellites left no doubt in either the East or West about Pretoria's intentions. Only a concerted diplomatic effort by leading Western and Eastern states helped to prevent the birth of yet another nuclear state. This cooperation between the United States and the Soviet Union, the United Kingdom, France and other states set a precedent in world policy on the containment of nuclear proliferation. Unfortunately it has not yet become a norm due to their considerable disagreement about the arms race, including who is ahead and who is behind. But as the increasing danger of proliferation during the period 1977–89 shows, without such cooperation, it will hardly be possible to maintain the non-proliferation regime.

The relaxed attitude of the United States towards South Africa on non-proliferation in the 1980s did not lead to the results the Reagan Administration counted on. Although Pretoria promised to follow the letter and spirit of the NPT in the export of its fissionable materials and nuclear technology, it categorically refuses to place its uranium enrichment plants under IAEA control. Pretoria plays its nuclear card whenever the United States and other Western countries seek to adopt economic sanctions against its policy of apartheid.

By the end of the eighties, the scenario in which South Africa's rulers would risk an open nuclear test to frighten the national liberation movement, raise the spirits of the white minority and "psychologically slap the leading Western nations in the face" for their unfriendly conduct, seems ever more real. Pretoria's quest for nuclear weapons has always been motivated more by political than military reasons. With its substantial military superiority in conventional forces over the front-line African states, South Africa has little to gain from nuclear weapons. They would be useless for internal riot control. As a "deterrent" to neighbouring countries and a means of upgrading South African political weight in the eyes of Western countries, the development and enhancement of its nuclear option has always been highly valued by Pretoria's ruling circles. At present, should the very existence of the white minority regime be jeopardized, its leaders can resort to the extreme measure of demonstrating its nuclear potential by conducting a nuclear test. Without doubt the absence of an international CTB would make it much easier for Pretoria to undertake such action. Nuclear states should take preventive action in the form of a comprehensive test ban.

Two Latin American threshold states, Argentina and Brazil, are also acquiring nuclear materials and technology that can be used for military purposes. Although the shift from military to civilian rule has weakened the military elite's influence on their nuclear programmes, the pace of these programmes has not abated. Moreover, Argentina and Brazil are beginning to cooperate in the nuclear field and to export nuclear materials and technology to the developing countries.

While branding the NPT as "discriminatory", Argentina and Brazil waste no effort in preserving freedom of choice in nuclear matters. They have both signed the Treaty of Tlatelolco and maintain their right to peaceful nuclear explosions, to enrich uranium and to extract plutonium from spent fuel, all without IAEA control. In addition, they are actively pursuing independent development of space and missile technology that can be applied to vehicles capable of delivering nuclear weapons.

Whether Argentina and Brazil develop nuclear weapons depends to a large extent on what happens in the international arms race. Will they feel the need to upgrade their political weight, by openly declaring themselves nuclear powers as did Britain, France and China? Their attitude is very similar to India's. Without making their choice final, Argentina and Brazil can take an intermediate step by conducting peaceful nuclear explosions without fear of their neighbour's reactions, because the Treaty of Tlatelolco permits such tests. This course might also be chosen to warn the nuclear states of the necessity to take into account the political interests of developing nations. It would be an especially seductive one for Argentina given its present stalemate with Britain over the Falkland (Malvinas) Islands.

It should be noted that Argentina and Brazil support a global test ban as the most effective step towards ending the nuclear arms race. Argentina, along with India, takes part in the meetings of the Six Nation Initiative calling for a ban on nuclear tests. Brasília tabled the proposal to declare the South Atlantic a zone of peace, free of nuclear weapons; Argentina signed, and Brazil signed and ratified, the 1964 Partial Test Ban Treaty. No doubt they would both join a CTB, and this willingness should be exploited to prevent their carrying out peaceful nuclear explosions.

In brief, a comprehensive test ban can be decisive on the threshold states' choice of the non-nuclear path. The great majority of these states would adjust their behaviour to match that of the leading nuclear powers.

Conclusion

A CTB would create a large political obstacle to the acquisition of nuclear weapons by those states which remain outside the NPT. They would find it extremely difficult from a political point of view to reject a CTB with worldwide support, especially since they could not characterize it as "discriminatory" as they did the NPT. This treaty, with identical obligations for both nuclear and non-nuclear states, would become an important integral part of the non-proliferation regime. Its signing by the states not party to the NPT would become a prelude to their joining it.

It is often said that even the threshold states' ratification of a CTB would not stop them from developing nuclear weapons. The possibility of conducting tests of nuclear devices in laboratories exists. But this is a real option only for those states with a high degree of technological sophistication. Only Israel fits in this category. Even for these states there comes a point, as was the case for the five declared nuclear states, when a nuclear test is needed to verify that an advanced weapon performs as designed. Here too a CTB would become an effective obstacle to a political decision to conduct such a test.

Thus, the realization of full-scale and effective nuclear weapon development programmes by the threshold states can be substantially slowed at an early stage. At the same time, a CTB's political influence on the threshold states' behaviour in the nuclear field would be impressive. A ban on nuclear testing under a CTB would be effectively the same as the strictures of the NPT are on those states which have the industrial base to produce nuclear weapons but choose not to. It is noteworthy to recall that President Kennedy's pessimistic scenario about the appearance of 15 – 20 nuclear states in the 1980s never materialized thanks only to the NPT and to the decision by the majority of threshold states party to it not to develop nuclear devices.

Even if some threshold states remain outside a CTB, its existence and support by the international community would be a factor they would have to contend with. It is well to recall here that the NPT's critics – the threshold states not party to it – had to come to terms with the prevailing non-proliferation policy of the world's nations. The conclusion of a CTB will allow the nuclear powers to take strong non-proliferation positions and give them more latitude to adopt joint or concurrent sanctions against states seeking to conduct nuclear tests.

At the end of the 1980s one of the main tasks facing us is finding a more effective means of ending the very real threat of horizontal proliferation. In a world where nuclear weapon technology is an open secret, the most effective way to prevent an enlargement of the nuclear club is to dissolve it. A CTB would signify the nuclear powers' resolve to do just that. The stakes in world non-proliferation policy are so high that neither the United States, nor the Soviet Union, nor any other nuclear power, can afford the "luxury" of testing newer and newer types of weapons for the continuation of a nuclear arms race that will ultimately boomerang on their vital national security interests.

References

1 *Arms Control and Disarmament Agreements: Texts and History of Negotiations* (Washington: US Arms Control and Disarmament Agency, 1977) pp. 84–86
2 UN Document A/C, 1/PV, 1556
3 Jimmy Carter, *Bulletin Atomic Scientists*, (October 1976) p. 11
4 *Disarmament*, vol. VIII (Winter 1985) no. 3, pp. 89, 95–97
5 *Fiscal year 1988: Arms Control Impact Statements, Statements Submitted to the Congress by the President*, (June 1987) p. 18
6 *Disarmament*, op. cit., p. 55
7 K. Subrahmanyam (ed.), *India and the Nuclear Challenge*, (New Delhi, 1986) p. 10
8 *Time*, (30 March 1987) p. 14
9 Leonard Spector, *Going Nuclear*, (Cambridge, MA: Ballinger, 1987) p. 230

International Actions to Stop and Reverse Proliferation of Nuclear Weapons

Theodore B. Taylor

Pending Actions Related to Non-Proliferation

The world faces a full agenda of urgent on-going or proposed actions designed to reduce sharply the short- and long-term likelihood of nuclear war. All the items on this agenda are significantly interconnected. My focus here is on some of the items of this agenda that relate especially to controlling the spread of nuclear weapons. These are listed below in a sequence of scheduled or possible times for their completion:

1. Initial rounds of unofficial comparative assessments of Gorbachev's proposal to eliminate all nuclear weapons (Zero Nuclear Arsenals, or ZNA) by the year 2000, Gandhi's proposal to do this no later than 2010, and various proposals for present Nuclear Weapon States (NWS) to maintain finite, but much smaller than present sized nuclear deterrent forces (Finite Nuclear Arsenals, or FNA) for the forseeable future (i. e. for at least the next three decades) – preliminary assessments by various, probably non-governmental groups, published before the end of 1989;
2. Convening of the first conference to consider amendments to the Partial Nuclear Test Ban Treaty (PTB) that would convert it into a Comprehensive Nuclear Test Ban Treaty (CTB) – possibly by April 1990, or a year after the formal requests, completed in April 1989, by the 39 parties to the PTB needed for such a conference to be required;
3. Continued negotiations of the START Treaty between the USA and USSR, calling for about 50 percent reduction in deliverable strategic nuclear warheads of each country – possibly in force by mid-1990;
4. Non-proliferation Treaty (NPT) Review Conference – August 1990;
5. Adoption or rejection of a CTB, as formulated by initial and later conferences of parties to the LTB – after mid-1990;
6. As yet unspecified actions related to binding commitments by the nuclear superpowers to achieve specified short- and long-term goals related to nuclear disarmament, especially as they relate to their responsibilities under Article 6 of the NPT – prior to 1995;
7. Conference of parties to the NPT to decide whether to prolong the Treaty indefinitely or for an additional fixed period or periods – early in 1995, if literally 25 years after the NPT went into force.

This is by no means a complete list of pending actions related to nuclear arms control and disarmament. Others include: steps towards treaties or moratoria on the production of nuclear-weapon materials; elimination of parts of or all stockpiles of short-range nuclear missiles in Europe; elimination of nuclear warheads (not called for in the INF treaty or current proposals for START); elimination of all tactical nuclear weapons; prohibition of any weapons in space; bans on chemical and biological weapons; sharp reductions in conventional weaponry, aimed at some sort of parity between potential or actual ememy countries; shifts in national priorities from military activities to intensive constructive action to deal with the multitude of severe environmental, economic, and other non-military threats facing humanity; international steps explicitly concerned with the eventual abolition of war as a legitimate human activity.

It is easy to add to this list. It is also easy to show that there can be significant connections between the needs for and ways to accomplish such actions. My intent here is to highlight a few of the many connections between the actions in the first of the above lists by considering each of them in sequence.

Assessments of Proposals for Worldwide Abolition of Nuclear Weapons

In spite of Gorbachev's call for the abolition of nuclear weapons by 2000, and Gandhi's call for their abolition and a global shift to a purely defensive military posture no later than 2010, I have been unable to find any published assessment, in some detail, of these proposals. By and large nuclear abolition proposals, even when made by heads of state of major powers, are being dismissed offhand as utopian and unrealistic by most arms control and disarmament experts.

I make this point here because of evident connections between any binding commitments to, or rejection of complete nuclear disarmament, especially by the superpowers, and the willingness of other countries to remain or to become parties to the NPT as it stands, or in a revised version.

As long as countries continue to behave as though they are more secure with nuclear weapons than without them, and show no clear evidence of seriously considering other alternatives, it is to be expected that other countries, which may or may not currently have secret or disclosed nuclear weapons, will follow suit.

I therefore strongly urge that all of us at this symposium press vigorously for careful and thorough assessments of these options, with publication of at least some substantive results before the start of the 1990 NPT Review Conference. Some of us are already involved in such an effort, but much more work is urgently needed. It would also be very helpful, to say the least, if the needed assessments were undertaken and promptly published by governmental enterprises representing established nuclear-weapon states, nuclear threshold states, and other countries.

A Comprehensive Nuclear Test Ban

Contrary to some views, a CTB (with no threshold yield below which nuclear explosions would be allowed) would block development of new and dangerous generations of nuclear weapons. This would hold not only for the superpowers, but also for threshold nuclear countries or any that may have already secretly crossed the threshold. Far from being only a symbolic gesture, establishment of a CTB would be a major element in the control of nuclear arms for those countries developing any but the most rudimentary types of nuclear explosives.

Pure fission nuclear explosives that are close to critical in their initial configurations before they are detonated can be developed with confidence without nuclear tests. Adherence to a CTB would, therefore, not prevent development and stockpiling of such weapons.

Fission explosives that use substantially less than one critical mass of fissile material at normal density (i.e. less than about 5 kg of plutonium or less than about 17 kg of uranium-235), or boosted or thermonuclear nuclear explosives, on the other hand, cannot be confidently developed without testing.

Unboosted fission weapons using much smaller than initially critical quantities of fissile materials can be confidently developed if "zero-yield" nuclear tests are allowed. These are tests with yields equivalent to very small quantities of high explosive (as little as a few grammes equivalent), still high enough to be measured accurately by close-in radiation measurements.

Very low yield nuclear tests (with yields less than 1 kiloton) can also provide very useful data related to the development of some new types of nuclear explosives. Examples include directional explosives that emit a significant fraction of their energy as ultra-high-velocity particles or droplets for use against targets in space or the upper atmosphere, or as bursts of microwaves for attacking targets in space or on the ground from explosions in either environment.

Successful development of inertial confinement fusion systems could provide very small nuclear explosions that could significantly assist the development of sophisticated thermonuclear explosives, or provide weapon-related data on the effects of nuclear explosives for new military applications.

It is for these reasons that many of us are pressing for a CTB with no yield threshold. An exception might be made for inertial confinement fusion experiments, but only under international safeguards to assure no secret withholding of the data or development of such systems for military purposes.

In short, the current LTB does not significantly inhibit either vertical or horizontal proliferation of nuclear explosives (except where above ground effects tests are needed), but a CTB would.

Current resistance by the United States and the United Kingdom to further restrictions on nuclear testing indicate that the most realistic prospect for a CTB is via the process of amendment of the PTB, the first step of which has already been taken.

Additional resistance to a CTB may also come from threshold countries that are quite willing to abide by the LTB, but for whom a CTB with no yield threshold would block development of all but the most rudimentary (and rather large and heavy) nuclear warheads. A hopeful sign on this score is that the only nations that voted against the 1986 UN resolution recommending steps towards convening the LTB amendment conference were the USA, UK, and France (not a party to the treaty). Israel and South Africa abstained. India, Pakistan, Brazil, and Argentina voted for the resolution.

In any case, prompt development of persuasive arguments for the USA, UK, France, and the threshold nations to support a CTB is called for if a veto of the needed amendment by the USA or UK is to be avoided. Special attention needs to be given to this matter during the next few months, before convening the conference in 1990.

START

Perhaps the greatest impact of START on proliferation will be to accelerate it, both vertically and horizontally, if the superpowers fail to negotiate a treaty soon. The hopes raised by the recent dramatic improvement in East-West relations, the coming into force of the INF Treaty, and widespread expectation that the START negotiations will be promptly concluded, have helped generate a situation in which failure to negotiate the treaty within the next year or so could cause the sudden collapse of various arms control regimes, including the NPT.

Some of us have argued that, unlike INF, provisions should be included in START for eliminating the relevant *warheads*, as well as the missiles. If this would substantially delay concluding a treaty, however, I would strongly urge concluding it without this provision, for the reason given above.

NPT Review Conference

The state of progress on a CTB is bound to be a major issue. That is why it is of such great importance that there be considerable substantive achievement at the first conference, probably several months earlier, to review proposed amendments of the PTB to convert it to a CTB. At the very least, it is to be hoped that this conference will have considered treaty language and associated verification protocols in sufficient detail to provide a basis for the NPT Review Conference to be able to support it as part of the action to be expected before the major review of the NPT in 1995.

Another set of issues that may be more prominent than in previous NPT review conferences involve the definitions of proliferation and of a nuclear weapon. There is practically a continuum of degrees of approach to the possession of nuclear weapons. Several countries have apparently progressed considerably along this continuum since the last NPT Review Conference. The non-nuclear parts of various kinds of nuclear explosive assembly systems can be developed and tested with high reliability without using any fissile materials, the principal items controlled under IAEA/NPT safeguards. How far can a country be expected to go in the design and non-nuclear tests of nuclear explosive assemblies, without violating the NPT? Would inertially confined, very small nuclear explosions used to power weapons producing pulses of electromagnetic radiation correspond to our present understanding of the meaning of the phrase "nuclear weapon"?

Published assessments of alternative nuclear disarmament and arms control objectives, including those listed above, would help considerably in assuring that the NPT Review Conference focuses on substantive issues related to future action prior to 1995, rather than dwelling on past failures to accomplish the principal global objectives of the NPT.

Adoption/Rejection of a CTB

According to currently expressed government policies, the Soviet Union and China are expected to favour a CTB promptly; the USA and UK are strongly opposed, and France is probably opposed. Assuming no big surprises in the formulation of a PTB treaty amendment converting it to a CTB, it seems reasonable to expect the needed majority of the parties to the PTB to vote for the amendment. Thus, a credible result of these actions would be for the USA and the UK, both parties to the Treaty, to use their option, under the Treaty, to veto the amendment. This is the reason for stressing here any actions that can help change the US position. Action that might help considerably would be votes for the treaty amendment by all the present threshold countries, all of which are parties to the PTB. The USA has consistently been a strong supporter of the NPT, and might change its policy towards a CTB if it sees that as buttressing its non-proliferation policies, while adherence to present testing policy stimulates accelerated proliferation and its attendant dangers.

Binding Commitments by the Superpowers to Massive Nuclear Disarmament

It is to be expected that both horizontal and vertical proliferation will continue without some further commitments by the superpowers to nuclear disarmament. Such commitments, even to total nuclear disarmament, may be *necessary* but are not likely to be *sufficient* to assure that all countries will agree to abolition of all nuclear weapons.

Comparative assessments of alternative global nuclear disarmament goals, including prompt achievement of abolition (ZNA) or minimum deterrence (FNA), therefore will require analysis of various levels of assurance of corresponding disarmament that might be provided by the superpowers, unilaterally ,bilaterally, or multilaterally. Is it reasonable to assume, for example, that some threshold countries may refrain from accepting the restraints imposed by the NPT until the goals of ZNA, or a form of FNA corresponding to cuts of current warhead stockpiles by more than 95 percent, for example, have actually been attained by the superpowers? If not, what lesser levels of commitment to either of these or other disarmament goals would be necessary for general acceptance of the NPT? What other commitments, by the superpowers and other countries (especially current adversaries), would be needed? Are superpower commitments to maintain nuclear forces to use in retaliation to attacks (nuclear or non-nuclear) on associated alliance members that do not have nuclear weapons necessary for all such members to adhere to the NPT?

Such questions are likely to be raised at the 1990 NPT Review Conference, and even more pointedly in the intervening time before the 1995 conference on the future of the NPT. Attempts to answer them, so far, have been inadequate.

1995 Conference on the Future of the NPT

Resolution of the future of the NPT in 1995 is called for by the Treaty itself. Some possible expectations of how this resolution may come out are likely to be discussed at the 1990 NPT Review Conference. Among the possibilities are:

1. Extension of the Treaty for some specified time in essentially the same form as it is now, keeping it open for other signatories; this might include an extension that is conditional on specified non-parties to the Treaty becoming parties by specified times;
2. Substantial amendment of the Treaty as it relates to controls of nuclear technology that prevent both horizontal and vertical proliferation, but not as it might relate directly to elimination of existing nuclear weapons (i. e. amendment to address what might be called "general and complete non-proliferation");
3. Further amendment of the Treaty to convert it to one covering the total abolition of nuclear weapons, perhaps with some temporary exceptions related to maintaining specified very low levels of minimum nuclear deterrence capacity by some countries or alliances;
4. Replacement of the NPT by another treaty calling not only for non-proliferation but also for international nuclear disarmament to zero, or to small but finite numbers of warheads;
5. Termination of the Treaty without any replacement.

These possibilities, and undoubtedly others, need to be assessed well before 1995. Some results, derived either by government or by non-government organizations, and published prior to the 1990 NPT Review Conference, could help assure substantive progress at that conference.

Using the results of these assessments, a detailed strategy for achieving the international consensus necessary for achieving the most desirable result needs to be developed and internationally discussed well before the 1995 review of the Treaty.

Beyond Non-Proliferation

Bhalchandra Udgaonkar

Some History

The first atomic weapons were exploded over Hiroshima and Nagasaki in August 1945, six weeks after the signing of the UN Charter. In its very first resolution, arising from the initiative of the five permanent members of the Security Council and Canada, adopted in January 1946, the General Assembly of the United Nations unanimously decided to establish an Atomic Energy Commission which was to make *specific proposals for the elimination of atomic weapons from national armaments*. The commission could not make any headway because of the irreconcilable differences between the superpowers. What survived its deliberations and related discussions at the United Nations and elsewhere, at the hands of the great powers (which one by one developed their own nuclear weapon capabilities in the meantime), was the idea of non-proliferation, decoupled from the disarmament dimension [1], in due course leading to the 1968 Treaty on Non-Proliferation of Nuclear Weapons. While much is made of the fact that the NPT has 140 parties, one should note that the non-signatories include not only two nuclear weapon powers, namely China and France, but also several threshold powers, and that 45 percent of the world's population is in states which have not accepted the NPT. Even among some of the signatory states there have been serious discussions of going nuclear after 1995, if the present nuclear-weapon states are allowed to retain a minimum deterrence.

In 1961, the same year that Ireland presented a draft resolution to the General Assembly that was the first of a series that would lead to the NPT [1], the non-aligned nations also moved a resolution in the General Assembly declaring that the use and threat of use of nuclear weapons would be a direct violation of the Charter of United Nations, and that any state using such weapons would be considered as acting contrary to the laws of humanity and as committing a crime against mankind and civilization. It demanded a total prohibition on the use of nuclear weapons. The resolution was then passed with a vote of 55 (including all the non-aligned countries) voting in favour, 20 voting against (most NATO countries and their allies) and 26 (including the socialist bloc countries) abstaining. In more recent years, while more than 136 member states of the UN have voted repeatedly calling for a ban on the use and the threat to use nuclear weapons, the resolution has been persistently opposed by the 15 NATO countries and two of their allies. Such persistent opposition underlines the fact that the NATO countries, inspite of having signed the NPT, consider nuclear weapons to be legitimate weapons. It also means that they do not take seriously the reminder in the preamble of the NPT, that in accordance with the Charter of the United Nations, states must refrain in their international relations from the threat or use of force against the territorial integrity or political independence of any state. All this is not calculated to inspire confidence in the NPT regime.

The approach of the non-aligned nations [2], from the very beginning, has been strongly against the legitimacy of nuclear weapons as instruments of policy, and correspondingly against any discrimination between the nuclear-weapon states and non-weapon states. In the matter of non-proliferation, the second non-aligned summit at Cairo (1964) called upon the great powers to abstain from all policies conducive to the dissemination of nuclear weapons among states that did not possess them, and at the same time to agree on measures providing for gradual liquidation of the existing stockpiles of nuclear weapons. The Conference also took an unambiguous stand against the deployment of nuclear weapons on the territories of non-nuclear states as one step towards the gradual liquidation of the stockpiles and denuclearization.

In 1965, India and seven other nations presented a resolution [3] to the General Assembly of the United Nations on a Treaty to Prevent the Proliferation of Nuclear Weapons, which called for negotiation of an international treaty to prevent the proliferation of nuclear weapons based on certain principles. These included the following:

(a) the treaty should be void of any loopholes which might permit nuclear or non-nuclear powers to proliferate, directly or indirectly, nuclear weapons in any form;
(b) the treaty should embody an acceptable balance of mutual responsibilities and obligations of the nuclear and non-nuclear powers;
(c) the treaty should be a step towards the achievement of general and complete disarmament and, more particularly, nuclear disarmament;
(d) there should be acceptable and workable provisions to ensure the effectiveness of the treaty.

This resolution was passed by the General Assembly with an overwhelming majority that included the UK, USA and USSR. The NPT which eventually emerged, however, did not accommodate the above concerns and principles. The elimination of nuclear weapons almost disappeared from the international agenda for the next twenty years, inspite of repeated reminders from the non-aligned movement [2] and the peace movements. The superpowers followed a path of mutually agreed armaments build-up rather than one of mutually agreed phased arms reduction, showing scant respect for their obligations under Article VI of the NPT. And the other three nuclear powers also built up their own nuclear arsenals as they saw fit.

The Present Situation

Vertical proliferation has gone on apace during the last two decades, unaffected by the NPT, and the number of nuclear warheads in the world has increased from less than 10000 at the time when the NPT was signed in 1968, to about 60000 today, with greater and greater sophistication in weapons and delivery systems thrown in. In relation to the euphoria unleashed by the INF agreement and the (so far unfulfilled) prospect of 50 percent cuts in strategic weapons, one must note that even if 85 percent of the present stockpile of nuclear warheads were to be dismantled, the world would still have more nuclear warheads then existed at the time of the signing of the NPT in 1968.

Testing of nuclear weapons and improvements in weapons have continued unabated. There has been no progress towards a Comprehensive Test Ban Treaty which was the declared intention of the signatories when the Partial Test Ban Treaty was signed in August 1963, an intention which was also recalled in the preamble to the NPT five years later. Even the 1974 Threshold Test Ban Treaty between USA and USSR, and the similar agreement signed between them in 1976 to limit the yield of peaceful nuclear explosions, have not been ratified.

One may remark here that signing a threshold test ban treaty at this stage will be counterproductive. It will clearly imply a fresh legitimization of the nuclear arms race, including the development of new nuclear weapons.

The Pugwash Council, in the Statement issued by it after the Annual Conference at Dagomys in September 1988, has ruefully noted:

"There still is more talk about disarmament than there is action, and deployments of new weapons systems are proceeding at a rapid rate". [4]

The Ignored Horizontal Proliferation

The number of declared nuclear-weapon states has long remained static at five. There has, however, been continued expansion, in both qualitative and quantitative terms, of nuclear weapon storage, command, control and communications facilities which constitute the nuclear war-fighting infrastructure, and now spread over at least 62 nations, principalities and territories [5]. There has also been an increase in the number and capabilities of nuclear-weapon-carrying submarines and other naval vessels in all oceans of the world, with the possible exception of the Antarctic. The horizontal proliferation that this implies, in terms of the number of independent, semi- and quasi-independent decision makers *within* the national structures of the five individual nuclear weapon powers, has not received as much attention in discussions of the NPT, as the possible horizontal proliferation to non-nuclear weapon states. Whatever pretence one may have of centralized political control of such a vast, world-wide, complex technological apparatus of nuclear weaponry in peace time, as Paul Bracken has vividly pointed out [6], in a time of crisis, the nuclear command systems in the nuclear weapons states, including USA and USSR, are likely to pass rapidly from political to highly fragmented military control. Jasjit Singh [7] has emphasized that the delegation of authority to such decision-makers is likely to increase in future as reaction-response times reduce, thus further increasing the number of independent decision-makers; and further, that during times of high alert decision-making was very likely to pass into the hands of so-called non-nuclear weapon allies too. *The real horizontal proliferation* has thus been within the nuclear weapon powers, to non-nuclear weapon states and territories, and also to crypto-nuclear weapon powers, whose nuclear weapon programmes have been aided, abetted and connived at by the nuclear weapons powers and their allies [8], violating Article I of the NPT.

Discussions of horizontal proliferation are usually confined to the possible acquisition of nuclear weapons (or capabilities) by Third World countries. Not only is similar concern not visible regarding possible acquisition of a nuclear weapon capability (and arsenal) by South Africa and Israel, but one finds hardly any discussion of the nuclear capabilities of Japan and many European countries and even of corporations, and what these imply for the stability of the NPT regime. Is this because the ethnocentric advocates of the NPT are not worried about horizontal proliferation among their allies, and are concerned only about a possible spread to Third World countries which are not part of their alliance systems?

In his paper at this Symposium, David Fischer [9] makes an interesting argument while discussing the nuclear weapon capability/arsenal of Israel:

"... in all circumstances it is vital that the USA should continue to persuade Israel to refrain from openly proclaiming her nuclear capability. An Israeli nuclear test or public declaration that she was a NWS would send a destabilizing shock-wave throughout the Middle East which could also have damaging consequences for the 1990 and 1995 Conferences and for the future of the NPT. It would face the USA with a painful dilemma. If the USA failed to react with sufficient vigour, her own non-proliferation policy would suffer a severe loss of coherence and credibility".

This argument hardly needs comment.

Global strategic and/or commercial considerations of nuclear-weapon states and their allies and of their corporations have often had precedence over the declared non-proliferation concerns of the big powers.

Doctrines and Attitudes

In the context of these developments, one has to examine the legitimizing role played by pernicious concepts and doctrines like nuclear deterrence, flexible response, limited nuclear war, etc., which are logically inconsistent with a non-proliferation policy (unless it is merely aimed at keeping the newly industrializing nations down), and yet have not been given up by certain vociferous advocates of the NPT. Correspondingly, many of the so-called non-nuclear-weapon countries, which are signatories of the NPT and also members of the NATO alliance, not only have nuclear missiles stationed on their soils, but also continue to train their troops in the use of nuclear weapons and have operational plans for their use. This is inconsistent with the intent and spirit of Articles I and II of the NPT.

The nuclear-weapon states have not appreciated the stand taken by the non-aligned nations against the discriminatory approach embodied in the NPT. The NPT of 1968 that they continue to promote, legitimized the possession of nuclear weapons in the hands of the five nuclear-weapon states, demanding at the same time that others renounce their nuclear option – thus dividing the world permanently into nuclear-weapon and non-nuclear-weapon states. The NPT has also been construed as permitting nuclear-weapon states to disseminate nuclear weapons through their deployment on others' territories, and even to bring pressures on tiny states to accept the stationing of nuclear weapons if the global strategy of a superpower requires it [10]. One also notes the recent pressure on FRG [11] to accept the modernization of short-range and tactical nuclear weapons as a part of NATO strategy, despite the mounting opposition to them in FRG.

Nuclear-weapon states and their allies have thus been going ahead with vertical proliferation and global horizontal proliferation, infesting the world's oceans with submarines and surface vessels carrying nuclear weapons (some of which made special appearances in times of crisis in the Indian Ocean zone and in the South Atlantic zone covered by the Treaty of Tlatelolco). If a littoral country with thousands of miles of coastline acquires or plans to acquire a nuclear propelled submarine (without nuclear weapons), this is described as being against the spirit of the NPT. In the meantime, it has not been possible to convene a Conference on the Indian Ocean as a Zone of Peace, ten years after the relevant Resolution was passed by the UN General Assembly.

Nuclear-weapon states have been advocating nuclear-weapon-free zones here and there, where it suits their global strategies. They have, however, not been willing to respect their sanctity by agreeing to keep their nuclear-weapon-carrying aircraft and naval vessels clear of these zones, or even to declare whether naval vessels carry nuclear weapons when they pass through such a zone or visit a port, or by agreeing not to conduct nuclear-weapon tests in such zones. In this context, one notes that while signing Protocol II of the Treaty of Tlatelolco, four of the five nuclear weapon powers have attached reservations which amount to reserving the option to use nuclear weapons against a member of the NWFZ. The Protocols of the Rarotonga Treaty for a NWFZ in the South Pacific have been rejected by three of the five nuclear-weapon powers, and further, one of the states in the zone covered by the Rarotonga Treaty, New Zealand, has had to face economic sanctions for deciding not to provide port facilities to ships which do not declare that they do not carry nuclear weapons. These facts cannot be lost sight of by other Third World countries which are being pressured to form other "nuclear-weapon-free" zones.

The attitude taken by the nuclear-weapon powers towards the rest has been well described by Sanders [1] as a taboo: "What you are doing is bad. I do it, and cannot stop doing it. But I shall do what I can to stop you".

Threat Perceptions and Confidence Building Between the North and the South

This attitude cannot but enhance the feeling of insecurity in the non-nuclear-weapon countries, in particular the Third World countries. Given the background of centuries of colonial domination and exploitation, and the current asymmetries in economic and technological strengths perpetuated through control of international financial institutions, including the IMF and the IBRD, and the inequitable terms of trade, including transactions in technology, the Third World countries are led to consider the asymmetry in nuclear capability in the NPT as yet another lever used by the rich industrialized countries to keep them under continued domination and control. The unilateral actions taken by them in restricting the supply of nuclear technology and materials (e. g. US Non-Proliferation Act of 1978; the formation of the Nuclear Suppliers' Club) are not only inconsistent with the intent of Article IV of the NPT and make the present (already discriminatory) NPT regime even more one-sided [12], but they also enhance the insecurity of the Third World in the critical area of energy.

Environmentalists have drawn attention to the increasing menace of the greenhouse effect, the need to lower the emission of CO_2 into the atmosphere, and, in this context to the need for a fresh look at the nuclear power option [13]. Nuclear power, through improved engineering designs to enhance safety, better management of radioactive wastes, and better institutional arrangements to prevent diversion of fissile material to weapons, may be expected to have an important role to play in keeping CO_2 emissions down in a world which requires substantially increased sources of power, especially in the Third World.

Such a large-scale expansion of nuclear power would, however, be difficult to achieve under the existing regime, with the international community divided into antagonistic blocs, with nuclear weapons deemed legitimate instruments of international power, and a discriminatory regime of technology transactions [14]. It is customary to recommend in the course of the NPT discussions that there should be no discrimination against any NPT members by suppliers in the provision of "exportable" nuclear technology, equipment and material, and that "an equitable code of conduct" for the transfer of technology should be agreed between suppliers and recepients. The existing embargos on commercial transactions in the non-nuclear area (even equipment needed for fundamental research in areas like cosmic rays or high-temperature superconductivity are subject to such embargos), and the stalled negotiations on the Code of Conduct for Technology Transfer at UNCTAD, do not inspire confidence in such pious recommendations.

Another disturbing trend noticed recently is that of questioning whether the obligations of the nuclear-weapons states under Article VI of the NPT were a quid-pro-quo for the obligations of non-nuclear-weapon states under Articles II and III, and of asserting that the obligations under Articles I, II and III (relating to horizontal proliferation in a narrow sense) are far more important than those under Article VI (relating to vertical proliferation) [15]. One has only to refer to the statements of President Lyndon Johnson, Prime Minister Harold Wilson, Premier Alexei Kosygin and others at the time of the NPT to recognize the disingenuity of such arguments. In any case, such arguments cannot but undermine the confidence of non-signatories and other non-aligned states in the intentions of the advocates of the NPT.

It is clear that the present NPT regime does not and cannot inspire confidence within the non-nuclear-weapon states, especially the Third World countries. While individually several non-aligned countries have agreed to join the NPT because of the coercive pressures from the nuclear-weapon countries, collectively they have refused to endorse it because of its discriminatory character. In the meantime, the coercive diplomacy of the great powers, which are also the nuclear-weapon powers, their interference and intervention in the affairs of others, continues. The concepts of confidence-building and common security, so assiduously built up in the East-West context, need to be extended on a global basis, in particular to North-South relations. New security doctrines have to be evolved [14] which look at common security on a truly global basis, and do not merely safeguard the interests of the great powers, or of the North.

The First and the Second World have yet to adjust themselves to the emergence of the Third World as a result of anti-colonial struggles leading to over a hundred newly independent nations, with their own identities and perceptions of global problems. These nations have their legitimate aspirations to make up for centuries of foreign rule, and wish to safeguard and consolidate their national independence so as to be able to embark upon a path of rapid development. They are dismayed when the doctrine of nuclear deterrence is given ethnocentric credit for forty "war-free" years when 17 million lives have been lost in over 150 wars since 1945, fought in their countries. The fact that half a dozen "regional" conflicts of long standing have recently started moving towards a resolution as soon as Presidents Gorbachov and Reagan arrived at a certain understanding, shows clearly, if indeed a proof was needed, that these "regional" conflicts were deeply embedded in the conflicts of the great powers in their quest for hegemony. The developing countries were equally dismayed when some of the big powers chose to keep away from the recent UN Conference on Disarmament and Development.

Need for New Thinking

In his address to the 38th Pugwash Conference at Dagomys, Professor Weisskopf [15] referred to the spirit of distrust which still dominated negotiations. "Ever-increasing deterrence is still the only way to "stabilize" world relations, a situation which will be considered in later times – if there will be people to consider it – *as rampant mental disease*", he said. One must confront this mental disease at its root, rather than attempting to treat symptoms, like proliferation, with psychological palliatives, like the NPT. In the same address, Weisskopf drew attention to three imminent global dangers: outbreak of a world war which would soon turn nuclear; the destruction of our environment; and thirdly, "upheavals in the under-developed parts of the world, where both power complexes are trying to expand their influence. Instead of helping these nations effectively to improve their living standards, most of the help is used to acquire weapons against competing groups". And he added, "The third danger may perhaps be the hardest to deal with. It is more than the others a political problem. *The transition from competition for influence to a collaboration for the improvement of the standard of living, requires a deep change of attitude, perhaps deeper than the change that may lead to an effective dealing with the other problems*". (emphasis added)

This is where new thinking is needed – more with regard to the North-South relations than the East-West relations, where there has already been a break-through, and with regard to an integrated approach to disarmament, development and environment.

New Pugwash Initiatives: Fourth NPT Review Conference

The need for such an integrated approach was recognized by Pugwash at Dagomys when it issued a special declaration entitled "Ensuring the Survival of Civilization" [16].

Dagomys also witnessed a beginning of some new thinking by Pugwash on the question of nuclear deterrence. Professor Rotblat [17] had recently argued that the permanent retention of nuclear weapons, even at a low level, is not acceptable for moral, political and strategic reasons, and had gone on to analyse ways to approach the goal of a nuclear-weapon-free world. The Statement of the Pugwash Council, issued at the end of the Dagomys Conference (September 1988) explicitly stated a nuclear-weapon-free world as a goal of Pugwash and also emphasized that "even under an arms control regime that included deep cuts in strategic forces and a CTB, the maintenance of any nuclear arsenal whatever, and of doctrines of nuclear deterrence, would pose dangers of nuclear weapons use as well as logical inconsistencies for non-proliferation policy".

The Pugwash Statement on the Fourth NPT Review Conference [18], issued from London on 30 November 1988, goes further: "Probably the most unfavourable factor of all" (vis-à-vis the NPT) says the Statement, "is the continued adherence of nuclear-weapon states to explicit or implicit doctrines that insist on the need for maintaining national nuclear forces as deterrents against nuclear or even conventional attack", and it adds: "It needs to be recognized that one of the gravest liabilities of nuclear-deterrence strategy, overt or covert, is its fundamental incompatibility with non-proliferation goals, in as much as *there is no logical basis for denying the "right" to a nuclear deterrent to some States while according it to others.* (emphasis added)

The London Statement [18] also:

1. recognizes the need to move ahead towards a more robust and more equitable non-proliferation regime, with decisive new actions;
2. expresses concern about the stability of the non-proliferation regime, *inter alia* from continuing vertical proliferation in the form of development and deployment of new nuclear weapon systems by the major powers, and recommends the adoption of *a comprehensive action plan with a specific time-table, for stopping and reversing the nuclear arms race* among the major nuclear powers and incorporating *inter alia*:
   - an early agreement on a prescription for including the other nuclear-weapon states in the build-down of nuclear arsenals;
   - an immediate moratorium on all nuclear testing, followed by the negotiation of a CTB Treaty;
   - a universal production cut-off on weapons-grade fissile material, under non-discriminatory international control;
   - rapid progress toward bringing under control the other dimensions of military completion and confrontation;
3. calls for a *formal commitment by all nuclear-weapon states not to be the first to use nuclear weapons* and not to structure their forces in ways that confer the appearance of first-use intention or capability, and not to threaten the use of nuclear weapons; and finally
4. calls for *steps to strengthen the NPT, in preparation for its extension in 1995,* beginning with an immediate initiation of a process, involving non-signatory states as well as signatories, aimed at agreement on means for enhancing the stability and equity of the non-proliferation regime.

What the recent initiatives of Pugwash imply is the recognition that one needs *a fundamental reappraisal of the* approach followed so far with regard to the problems of the nuclear arms race and disarmament. The 1990 NPT Review Conference will fail in its purpose if it does not face up to the questions discussed above and the recommendations of Pugwash.

In fact, one may go further and suggest a few more items explicitly for inclusion in a time-bound action-plan, so as to build up confidence among the non-nuclear-weapon states:

1. an agreement by the USA and USSR to place the explosive nuclear material (enriched uranium, plutonium and tritium), released as a result of the INF agreement and (eventual) agreement on deep cuts, in the custody of the IAEA;

2. an agreement by nuclear-weapon states, on a time-bound programme, to withdraw their nuclear weapons and C^3I infrastructures from the territories of non-nuclear states, as one concrete step towards the gradual liquidation of stockpiles and denuclearization;
3. an agreement by nuclear-weapon states to withdraw their nuclear weapons from aircraft, submarines and other naval vessels, and to allow on-board challenge inspections with regard to nuclear weapons;
4. an agreement to keep outer space free of all weapons, and a moratorium on all activities of developing, testing and deploying space weapon systems.

William Epstein [19] reminds us of the main objective of the NPT Review Conferences: "In order to make sure that the nuclear powers were living up to their obligations, the treaty provides for a review conference at intervals of five years".

The NPT Review Conference

The NPT is not an end in itself, but an instrument that was meant to lead to a cessation of the nuclear arms race at an early date and to nuclear disarmament and a treaty on general and complete disarmament under strict and effective international control (Article VI of the NPT). The objective was, and has to continue to be, a world from which national arsenals of nuclear weapons and all other major weapons of mass destruction, including the means of their delivery, have been eliminated. This objective, embodied in the very first resolution of the United Nations, unanimously adopted by the General Assembly in January 1946 (and repeated in the Preamble to the NPT), has not only remained unfulfilled, it has not even been taken up for serious negotiations by the nuclear-weapon powers.

This objective must be constantly kept in view as the focal point of all discussions of the NPT.

The preparations for the Fourth NPT Review Conference should not, therefore, narrowly confine themselves to the NPT regime as it exists today. *They must be broadened in scope so that the 1990 NPT Review Conference starts a process of reviewing the NPT itself*, and sets in motion a series of negotiations which could lead to the inclusion of the many important elements in the London Statement of Pugwash, so that a strengthened non-proliferation regime may emerge by the time of the 1995 Conference. Such a strengthened non-proliferation regime would have to be a comprehensive system of international peace and security, in which all countries participate on an equal footing. And it would have to be centred on the goal of achieving a nuclear-weapon-free world following a comprehensive action plan with a specific time-table, as emphasized in the London Statement of Pugwash. Further, it will have to include a multilateral mechanism for monitoring the implementation of its various stages.

All states, signatories and non-signatories of the NPT, will have to be a part of this process. One could draw on the experience of the Law of the Sea Conference to devise an appropriate methodology.

One is reminded that the Pugwash Statement on the Third NPT Review Conference, issued in 1985, had recommended that the Third NPT Review Conference should begin to concern itself with the post-1995 period, and "set in motion the preparation of a new formulation of the NPT for discussion at the Review Conference in 1990". Unfortunately, there has been no movement in this direction.

Beyond Narrow Non-Proliferation to a World Order Without Nuclear Weapons and War

It has become increasingly clear that the world has to move from the inherently discriminatory doctrine of non-proliferation, which seeks to perpetuate the present unjust and unstable world order, to the objective of a nuclear-weapon-free world, and to do so against a well-defined time frame.

The Russell-Einstein Manifesto has been the credo of Pugwash, but the objective of abolition of war which was projected in the Manifesto has been lost sight of even in Pugwash. It would be useful at the present juncture to remind ourselves of the problem posed in the Russell-Einstein Manifesto:

"Here, then is the problem which we present to you, stark, and dreadful, and inescapable. Shall we put an end to the human race; or shall mankind renounce war? People will not face this alternative because it is so difficult to abolish war. The abolition of war will demand distasteful limitations of national sovereignty ..."

The world community has to face up to such limitations of national sovereignty [20], if peace is to be stable, durable and *global*. One sees beginnings of movements in this direction in the Six-Nation Initiative and in the Delhi Declaration on Principles for a Nuclear-Weapon-Free and Non-Violent World [21], signed by Prime Minister Rajiv Gandhi and General Secretary Mikhail Gorbachev on 27 November 1986. These have to be pursued vigorously.

While discussing the difficulties with the complete elimination of nuclear weapons, Joseph Rotblat [17] has dealt with the concern as to how to ensure that one side will not be cheating in the last stages of nuclear disarmament, and conceal a small number of such weapons with which it could gain political power, if not world domimnation. In this context, he suggests the setting up of an international authority endowed with a small nuclear weapon capability. "The United Nations", he points out, "has proved itself in a variety of policing duties and peace negotiations, and has acquired enough status to be entrusted with an even mightier task, to be the custodian of an instrument designated to prevent a threat to world security". He also points out that "the treaty abolishing national nuclear arsenals should be preceded by the setting up of (this) international depository of nuclear weapons and means of delivery, sufficient to deter a potential violator of the treaty".

The world will need other new international instruments. A world without nuclear weapons will not be more safe than the world in 1939, unless we have adequate instruments for conflict resolution and peace keeping. The United Nations will, therefore, have to be suitably strengthened so as to be able to meet any threat of aggression with a standing international peace force. The Parliamentarians Global Action have made several other recommendations [22] which deserve serious consideration: e. g. an expanded system of world courts and institutions for mediation and arbitration, as an alternative to the battle-field as a way of settling disputes; and a global multilateral verification system for monitoring the gradual demilitarization of nations. As they point out, while more work needs to be done on the detailed problems of arms reduction, peacekeeping and enforcement, verification and conversion to a peacetime economy, the key elements were already outlined in a series of governmental and intergovernmental documents since the Second World War, including the principles of a comprehensive disarmament treaty agreed to by the United States and the USSR in 1961 (the McCloy-Zorin Principles), the USA and the Soviet draft disarmament treaties of 1962, and the Final Draft Document of the UN Special Session on Disarmament in 1978. These could be the starting point of fresh negotiations, involving all nations. And such negotiations will have to start soon, in parallel with the negotiations on the comprehensive action plan with a specific time-table for stopping and reversing the nuclear arms race.

Without such measures, considered conjointly with serious efforts towards a just world order, any peace or disarmament agreement is bound to be fragile. Durable peace and global common security demand an equitable and just world order. The fight for such a world order has to find a place on the agenda of any genuine efforts toward disarmament, peace and global security.

References

1 For a brief history of this early period see Ben Sanders (this volume p. 80)

2 For details see the chapter "The Non-Aligned Movement and Disarmament", in *Surviving the Second Millenium: Disarmament and Information*, research monograph for UNESCO, (New Delhi: Namedia Foundation, 1987). For a brief summary see B. M. Udgaonkar, "Non-Aligned Movement and Pugwash", *Special Supplement to New Times*, Moscow, August 1988

3 Resolution 2028 (XX) adopted by the UN General Assembly on 19 November 1965

4 *Pugwash Newsletter*, vol. 26, (1988), pp. 34–39

5 William Arkin and Richard Fieldhouse, *The Nuclear Battlefields – Global Links in the Arms Race*, (Cambridge, MA: Ballinger, 1985)

6 Paul Bracken, *The Command and Control of Nuclear Forces*, (New Haven: Yale University Press, 1983)

7 Jasjit Singh, "Nuclear Arms Race: Dimensions and Trends", paper for the conference *Towards a Nuclear-Weapon-Free and Non-Violent World*, (New Delhi, 14–16 November 1988)

8 See Leonard Spector, "New Players in the Nuclear Game", *Bulletin Atomic Scientists*, (January-February 1989) pp. 12–15 , for a recent review of this aspect, of how the global strategy of a nuclear weapon power over-rides its non-proliferation concerns. See also Richard Bolt, "Plutonium for All: Leaks in Global Safeguards", *ibid*, (December 1988)

9 David Fischer (this volume p. 36)

10 For a description of the problem faced by the tiny coral reef island of Palau in the South Central Pacific, population 15,000, see Jennifer Scarlott, *Bulletin Atomic Scientists*, (November 1988) pp. 31–35

11 Sunil Adam, "Nuclear Missiles: Bonn's Allergy", *Times of India*, Bombay, 26 April 1989

12 V. Knapp, "NPT Can and Should be Strengthened", Proceedings of the 38th Pugwash Conference, Dagomys, USSR, 1988

13 For example, the report of the Energy Working Group of the conference *The Changing Atmosphere: Implications for Global Security*, (Toronto, 27–30 June 1988)

14 K. Subrahmanyam, "Alternative Security Doctrines", paper for the conference *Towards a Nuclear-Weapon-Free and Non-Violent World*, (New Delhi, 14–16 November 1988)

15 Victor Weisskopf, *Pugwash Newsletter*, vol. 26, (1988) pp. 51–53

16 "Ensuring the Survival of Civilization", Dagomys Declaration of the Pugwash Council, *Pugwash Newsletter*, vol. 26, (1988) pp. 39–40

17 J. Rotblat, "The Elimination of Nuclear Weapons: Is it Desirable? Is it Possible"?, (in course of publication)

18 Pugwash Statement on the Forthcoming Fourth NPT Review Conference, *Pugwash Newsletter*, vol. 26, (1989) pp. 181–182

19 William Epstein, "The Proliferation of Nuclear Weapons and Efforts to Prevent Further Proliferation", *Proceedings of the Thirty-Seventh Pugwash Conference on Science and World Affairs*, (Gmunden am Traunsee, Austria, 1–6 September 1987)

20 It is the great powers that really have to gear up for the "distasteful limitations of national sovereignty" referred to in the Russell-Einstein Manifesto. For other countries such limitations are part of life

21 Delhi Declaration on Principles for a Nuclear Weapon Free and Non-Violent World, November 27, 1986

22 "Global Security", Parliamentarians Global Action, presented to the conference *Towards a Nuclear Weapon Free and Non-Violent World*, (New Delhi, 14–16 November 1988)

The Indian Proposal at UNSSOD-III: Problems and Prospects

Valerii Davydov

The prevention of nuclear proliferation is sometimes called the "second front" of the struggle against the nuclear threat. While some progress – including the INF Treaty and the ongoing negotiations for fifty percent cuts in American and Soviet strategic forces – has been achieved on the "first front" of the nuclear arms race by the five nuclear powers – USSR, USA, UK, France and China – the situation on the "second front" has not only worsened, it is out of control.

At present there are about 140 states party to the Treaty on the Non-Proliferation of Nuclear Weapons (NPT) that imposes the political obligation not to create such weapons. At the same time about thirty states still remain outside the framework of the Treaty, including two nuclear states – China and France – and a number of countries having well-developed nuclear industries – Argentina, Brazil, Israel, India, Pakistan and South Africa.

According to the Carnegie Fund's experts' estimates, Israel may already have a nuclear potential of 200 weapons, including thermonuclear ones. South Africa has the potential to produce about twenty nuclear weapons. By 1991 Pakistan might be capable of producing fifteen and India more than 100 nuclear weapons with yields equal to the Hiroshima bomb [1]. Argentina and Brazil are capable of producing nuclear weapons at short notice. A growing interest in the acquisition of nuclear technology and materials that might be used for military purposes is being shown by a large number of states party to the Treaty.

At present a new dimension of the nuclear non-proliferation problem – a missile one – is becoming highly visible. All of the strongest candidates to the "nuclear club" are wasting no effort in their attempt to develop indigenously, or to acquire, missile delivery vehicles.

If the development of missile delivery vehicles was previously considered to be very expensive both in time and finance, the present worldwide spread of missile technology presents an easier and faster solution to this problem. Moreover, the availability of ballistic missile technology might provide an incentive for nuclear weapons acquisition.

Some experts, both in the East and in the West, believe that if nuclear weapons are ever used again this most likely would happen in a conflict between states capable of producing nuclear weapons. Taking into account the political and military ties many countries have with the United States and the Soviet Union, the probability of them being drawn into local nuclear conflicts against their will is very high. Should the proliferation of nuclear weapons continue, the threat of nuclear terrorism – the anonymous use of nuclear weapons – will certainly rise, and a multiple and unpredictable growth of the nuclear threat as a whole will take place. The possibility of dismantling nuclear weapons might become unrealistic and unattainable.

All this should make nations cooperate at an unprecedented level in search of novel approaches to nuclear non-proliferation. By the early 1990s a new wave of nuclear proliferation will demand measures similar in scale and character to those undertaken in the 1960s when the NPT was signed.

The Indian proposal at UNSSOD-III should be examined against this background. According to it, in 1995 when the first term of the NPT expires and its member-states are to decide the future of the Treaty, international negotiations should begin on a new treaty to replace the NPT that would commit nuclear-weapon states to eliminate all nuclear weapons by the year 2010, and all non-nuclear-weapon states not to cross the nuclear-weapon threshold. The Indian proposal was controversial. On the other hand, the proposal was well-received by many developing nations, while, on the other major nuclear powers and their allies ignored it. Many nuclear experts were taken by surprise at the proposal, because it pointed towards a radical restructuring of relations between nuclear and non-nuclear states.

India's policy towards nuclear weapons, its negative attitude towards the NPT and its strong support of a CTB have been discussed earlier. [2]

Lacking a positive response to the Indian initiative, it is hardly possible to count on Indian acceptance of the nuclear powers' attempts to lecture or to dictate to it that the possession of a nuclear arsenal is bad and that it should forever forgo freedom of action in the nuclear field. Moreover, such an approach could accelerate India's nuclear activity.

After all, why should such states as India, Argentina, or Brazil not hope to acquire the status of great nuclear powers like the permanent members of the Security Council – Britain, France and China? It is another matter entirely as to whether they will choose to do this by the open possession of nuclear weapons.

The more these states earn political dividends from their non-nuclear status the less attractive would be the "nuclear choice". The paradox is that in order to stop the drive of the majority of threshold states to nuclear status they should be treated as de-facto nuclear powers. Most measures to restrain nuclear proliferation in South Asia, the Middle East, Africa, or Latin America can be accepted by threshold countries if they assume not only a regional but a global character as well:

- The firm refusal of these states to engage in nuclear testing is possible only under a comprehensive test ban treaty;
- The cessation of production of fissionable materials for military purposes in these states is attainable on condition of a moratorium for their production in the USA and the USSR;
- International inspection of all nuclear facilities is possible only in the case of admission to corresponding facilities of the nuclear powers.

In the final analysis, in exchange for slowing down nuclear preparations, India, Argentina, Brazil and other threshold states hope to earn essential political dividends by proving their ability to exert influence in the field of nuclear disarmament. The main motives behind the Indian proposal should be considered above all in this context.

It should be presumed that the majority of candidates to the nuclear club will support the Indian proposal and take part in its realization. This plan opens for them a shorter and less expensive route to the enhancement of their political prestige than the long and expensive one leading to the possession of a nuclear arsenal. At the same time the main aim of the international policy on nuclear non-proliferation is to obtain the political commitment not to "go nuclear" in treaty form.

Of course the Indian proposal does raise an important question: Should all the nuclear facilities producing fissionable material in threshold states be brought under IAEA safeguards? It seems that this task could be easily solved if the USA and other nuclear powers showed the same readiness as the USSR to abolish the production of enriched uranium and plutonium for military purposes. In any case, IAEA safeguards are essential for the verification of political commitments not to produce nuclear weapons.

On the whole, the Indian proposal, if accepted by all the threshold states, looks very promising, especially when there are no other effective political measures to prevent nuclear proliferation.

The Indian proposal is responsive to the political intentions of most participants of the NPT. Despite the steady growth of the number of its adherents, there is no certainty that all participants in the 1995 Conference will follow a scenario suitable to the nuclear powers: that the Treaty will be extended or made permanent. Such lack of confidence is mainly due to the divergence of opinion between nuclear and non-nuclear states participating in this Treaty as to its value and effectiveness. These contradictions are first of all connected with the problem of the mutual fulfillment of the Treaty's obligations.

At the present time it is difficult to say whether the non-nuclear states will support the Indian proposal or whether they will put forward some other proposals calling for inclusion in the Treaty's text obligations to ban nuclear testing completely, to cease producing fissionable materials for military purposes, to prohibit using nuclear weapons, and so on. It is evident that in order to avoid a crisis at the 1995 conference it is necessary for the nuclear powers to take forward-looking and far-reaching actions in the field of disarmament.

The Soviet programme to liquidate nuclear weapons by 2000, its efforts to ban nuclear testing and to stop the production of fissionable materials proves its readiness to go to the limit in the field of nuclear disarmament. However, such a readiness has not been evident up to now on the part of the USA and UK.

After the Soviet-American Summit in Reykjavík, Washington retreated from its proposals to create a nuclear-free world. It appears that the Bush Administration is not going to part with nuclear weapons. However, such a position by Washington is very fragile, because it tries to combine incompatible aims: to preserve for itself a free hand in nuclear preparations and at the same time to stop them in the threshold states. The alternatives before the United States are evident: to take far-reaching steps in nuclear disarmament or to reconcile itself to the appearance of new nuclear powers. Nuclear arrogance, dismissive of the opinion of non-nuclear states, renders bad service in solving this problem.

The same approach is characteristic of Britain. British praise of nuclear weapons as a tool of peace strengthens those circles in Israel, South Africa, India and Pakistan that call for the production of nuclear weapons. One does not have to be a prophet to imagine that with such an approach the USA and UK face hard times in 1995.

In principle, if the United States and Great Britain would agree to consider the Indian proposal, the question of participation in a new treaty by other nuclear powers – France and China – might be raised. France and China, non-members of the NPT, took on themselves an obligation to carry out its terms. These powers display a sincere interest in the reduction of the nuclear potential of the USA and USSR. They promise under certain conditions to join the process of eliminating nuclear arms.

It should be noted here that if the Indian proposal were accepted by the five nuclear powers as the framework of a new treaty, the universal obligations for nuclear disarmament could be defined for the first time. Sooner or later, depending on the results of Soviet-American negotiations on nuclear arms reduction, the necessity of such obligations will be put on the agenda of world politics. In the end, whatever the achievements of the NPT, it was always regarded as an intermediate step towards the highest aim – the liquidation of nuclear weapons.

The Indian proposal at UNSSOD-III will evidently be at the centre of attention of the world community. Its further fate will be determined not so much by the nuclear powers as by the support of the non-nuclear states. It is difficult to agree with the view that, in case of acceptance of the Indian proposal by most of the non-nuclear states, the world community will be left without any NPT at all. The majority of non-nuclear states participate in the Treaty, not because they were forced to do so, but because they understood that without nuclear arms their security was strengthened, and because they have in this Treaty a tool to influence the world arms race. If anything can undermine the NPT it is the policy of the nuclear arms race, which ignores the vital interests of the majority of world states, leading to the perpetuation of the division between nuclear and non-nuclear states. The future vitality of the NPT depends first of all on the readiness of the nuclear powers to remain on the road to the complete elimination of nuclear weapons.

References

1 Leonard Spector, *The Undeclared Bomb*, (Cambridge, MA: Ballinger, 1988) pp.70, 181, 293
2 Vitalii Goldanskii and Valerii Davydov (this volume p. 147)

Appendix: Treaty on the Non-Proliferation of Nuclear Weapons (NPT)

Signed at London, Moscow and Washington on 1 July 1968. Entered into force on 5 March 1970. Depositaries: UK, US and Soviet governments

The States concluding this Treaty, hereinafter referred to as the "Parties to the Treaty",

Considering the devastation that would be visited upon all mankind by a nuclear war and the consequent need to make every effort to avert the danger of such a war and to take measures to safeguard the security of peoples,

Believing that the proliferation of nuclear weapons would seriously enhance the danger of nuclear war,

In conformity with resolutions of the United Nations General Assembly calling for the conclusion of an agreement on the prevention of wider dissemination of nuclear weapons,

Undertaking to co-operate in facilitating the application of International Atomic Energy Agency safeguards on peaceful nuclear activities,

Expressing their support for research, development and other efforts to further the application, within the framework of the International Atomic Energy Agency safeguards system, of the principle of safeguarding effectively the flow of source and special fissionable materials by use of instruments and other techniques at certain strategic points,

Affirming the principle that the benefits of peaceful applications of nuclear technology, including any technological by-products which may be derived by nuclear-weapon States from the development of nuclear explosive devices, should be available for peaceful purposes to all Parties to the Treaty, whether nuclear-weapon or non-nuclear-weapon States,

Convinced that, in furtherance of this principle, all Parties to the Treaty are entitled to participate in the fullest possible exchange of scientific information for, and to contribute alone or in co-operation with other States to, the further development of the applications of atomic energy for peaceful purposes,

Declaring their intention to achieve at the earliest possible date the cessation of the nuclear arms race and to undertake effective measures in the direction of nuclear disarmament,

Urging the co-operation of all States in the attainment of this objective,

---

Source: *Treaty Series*, Vol. 729 (United Nations, New York).

Recalling the determination expressed by the Parties to the 1963 Treaty banning nuclear weapon tests in the atmosphere, in outer space and under water in its Preamble to seek to achieve the discontinuance of all test explosions of nuclear weapons for all time and to continue negotiations to this end,

Desiring to further the easing of international tension and the strengthening of trust between States in order to facilitate the cessation of the manufacture of nuclear weapons, the liquidation of all their existing stockpiles, and the elimination from national arsenals of nuclear weapons and the means of their delivery pursuant to a Treaty on general and complete disarmament under strict and effective international control,

Recalling that, in accordance with the Charter of the United Nations, States must refrain in their international relations from the threat or use of force against the territorial integrity or political independence of any State, or in any other manner inconsistent with the Purposes of the United Nations, and that the establishment and maintenance of international peace and security are to be promoted with the least diversion for armaments of the world's human and economic resources,

Have agreed as follows:

Article I

Each nuclear-weapon State Party to the Treaty undertakes not to transfer to any recipient whatsoever nuclear weapons or other nuclear explosive devices or control over such weapons or explosive devices directly, or indirectly; and not in any way to assist, encourage, or induce any non-nuclear-weapon State to manufacture or otherwise acquire nuclear weapons or other nuclear explosive devices, or control over such weapons or explosive devices.

Article II

Each non-nuclear-weapon State Party to the Treaty undertakes not to receive the transfer from any transferor whatsoever of nuclear weapons or other nuclear explosive devices or of control over such weapons or explosive devices directly, or indirectly; not to manufacture or otherwise acquire nuclear weapons or other nuclear explosive devices; and not to seek or receive any assistance in the manufacture of nuclear weapons or other nuclear explosive devices.

Article III

1. Each non-nuclear-weapon State Party to the Treaty undertakes to accept safeguards, as set forth in an agreement to be negotiated and concluded with the International Atomic Energy Agency in accordance with the Statute of the International Atomic Energy Agency and the Agency's safeguards system, for the

exclusive purpose of verification of the fulfilment of its obligations assumed under this Treaty with a view to preventing diversion of nuclear energy from peaceful uses to nuclear weapons or other nuclear explosive devices. Procedures for the safeguards required by this Article shall be followed with respect to source or special fissionable material whether it is being produced, processed or used in any principal nuclear facility or is outside any such facility. The safeguards required by this Article shall be applied on all source or special fissionable material in all peaceful nuclear activities within the territory of such State, under its jurisdiction, or carried out under its control anywhere.

2. Each State Party to the Treaty undertakes not to provide: (a) source or special fissionable material, or (b) equipment or material especially designed or prepared for the processing, use or production of special fissionable material, to any non-nuclear-weapon State for peaceful purposes, unless the source or special fissionable material shall be subject to the safeguards required by this Article.

3. The safeguards required by this Article shall be implemented in a manner designed to comply with Article IV of this Treaty, and to avoid hampering the economic or technological development of the Parties or international co-operation in the field of peaceful nuclear activities, including the international exchange of nuclear material and equipment for the processing, use or production of nuclear material for peaceful purposes in accordance with the provisions of this Article and the principle of safeguarding set forth in the Preamble of the Treaty.

4. Non-nuclear-weapon States Party to the Treaty shall conclude agreements with the International Atomic Energy Agency to meet the requirements of this Article either individually or together with other States in accordance with the Statute of the International Atomic Energy Agency. Negotiation of such agreements shall commence within 180 days from the original entry into force of this Treaty. For States depositing their instruments of ratification or accession after the 180-day period, negotiation of such agreements shall commence not later than the date of such deposit. Such agreements shall enter into force not later than eighteen months after the date of initiation of negotiations.

Article IV

1. Nothing in the Treaty shall be interpreted as affecting the inalienable right of all the Parties to the Treaty to develop research, production and use of nuclear energy for peaceful purposes without discrimination and in conformity with Articles I and II of this Treaty.

2. All the Parties to the Treaty undertake to facilitate, and have the right to participate in, the fullest possible exchange of equipment, materials and scientific and technological information for the peaceful uses of nuclear energy. Parties to the Treaty in a position to do so shall also co-operate in contributing alone or together with other States or international organizations to the further development of the applications of nuclear energy for peaceful purposes, especially in the ter-

ritories of non-nuclear-weapon States Party to the Treaty, with due consideration for the needs of the developing areas of the world.

Article V

Each Party to the Treaty undertakes to take appropriate measures to ensure that, in accordance with this Treaty, under appropriate international observation and through appropriate international procedures, potential benefits from any peaceful applications of nuclear explosions will be made available to non-nuclear-weapon States Party to the Treaty on a non-discriminatory basis and that the charge to such Parties for the explosive devices used will be as low as possible and exclude any charge for research and development. Non-nuclear-weapon States Party to the Treaty shall be able to obtain such benefits, pursuant to a special international agreement or agreements, through an appropriate international body with adequate representation of non-nuclear-weapon States. Negotiations on this subject shall commence as soon as possible after the Treaty enters into force. Non-nuclear-weapon States Party to the Treaty so desiring may also obtain such benefits pursuant to bilateral agreements.

Article VI

Each of the Parties to the Treaty undertakes to pursue negotiations in good faith on effective measures relating to cessation of the nuclear arms race at an early date and to nuclear disarmament, and on a treaty on general and complete disarmament under strict and effective international control.

Article VII

Nothing in this Treaty affects the right of any group of States to conclude regional treaties in order to assure the total absence of nuclear weapons in their respective territories.

Article VIII

1. Any Party to the Treaty may propose amendments to this Treaty. The text of any proposed amendment shall be submitted to the Depositary Governments which shall circulate it to all Parties to the Treaty. Thereupon, if requested to do so by one-third or more of the Parties to the Treaty, the Depositary Governments shall convene a conference, to which they shall invite all the Parties to the Treaty, to consider such an amendment.

2. Any amendment to this Treaty must be approved by a majority of the votes of all the Parties to the Treaty, including the votes of all nuclear-weapon

States Party to the Treaty and all other Parties which, on the date the amendment is circulated, are members of the Board of Governors of the International Atomic Energy Agency. The amendment shall enter into force for each Party that deposits its instrument of ratification of the amendment upon the deposit of such instruments of ratification by a majority of all the Parties, including the instruments of ratification of all nuclear-weapon States Party to the Treaty and all other Parties which, on the date the amendment is circulated, are members of the Board of Governors of the International Atomic Energy Agency. Thereafter, it shall enter into force for any other Party upon the deposit of its instrument of ratification of the amendment.

3. Five years after the entry into force of this Treaty, a conference of Parties to the Treaty shall be held in Geneva, Switzerland, in order to review the operation of this Treaty with a view to assuring that the purposes of the Preamble and the provisions of the Treaty are being realised. At intervals of five years thereafter, a majority of the Parties to the Treaty may obtain, by submitting a proposal to this effect to the Depositary Governments, the convening of further conferences with the same objective of reviewing the operation of the Treaty.

Article IX

1. This Treaty shall be open to all States for signature. Any State which does not sign the Treaty before its entry into force in accordance with paragraph 3 of this Article may accede to it at any time.

2. This Treaty shall be subject to ratification by signatory States. Instruments of ratification and instruments of accession shall be deposited with the Governments of the United Kingdom of Great Britain and Northern Ireland, the Union of Soviet Socialist Republics and the United States of America, which are hereby designated the Depositary Governments.

3. This Treaty shall enter into force after its ratification by the States, the Governments of which are designated Depositaries of the Treaty, and forty other States signatory to this Treaty and the deposit of their instruments of ratification. For the purposes of this Treaty, a nuclear-weapon State is one which has manufactured and exploded a nuclear weapon or other nuclear explosive device prior to 1 January, 1967.

4. For States whose instruments of ratification or accession are deposited subsequent to the entry into force of this Treaty, it shall enter into force on the date of the deposit of their instruments of ratification or accession.

5. The Depositary Governments shall promptly inform all signatory and acceding States of the date of each signature, the date of deposit of each instrument of ratification or of accession, the date of the entry into force of this Treaty, and the date of receipt of any requests for convening a conference or other notices.

6. This Treaty shall be registered by the Depositary Governments pursuant to Article 102 of the Charter of the United Nations.

Article X

1. Each Party shall in exercising its national sovereignty have the right to withdraw from the Treaty if it decides that extraordinary events, related to the subject matter of this Treaty, have jeopardized the supreme interests of its country. It shall give notice of such withdrawal to all other Parties to the Treaty and to the United Nations Security Council three months in advance. Such notice shall include a statement of the extraordinary events it regards as having jeopardized its supreme interests.

2. Twenty-five years after the entry into force of the Treaty, a conference shall be convened to decide whether the Treaty shall continue in force indefinitely, or shall be extended for an additional fixed period or periods. This decision shall be taken by a majority of the Parties to the Treaty.

Article XI

This Treaty, the English, Russian, French, Spanish and Chinese texts of which are equally authentic, shall be deposited in the archives of the Depositary Governments. Duly certified copies of this Treaty shall be transmitted by the Depositary Governments to the Governments of the signatory and acceding States.

Subject Index

FSC
www.fsc.org
MIX
Papier aus verantwortungsvollen Quellen
Paper from responsible sources
FSC® C105338

If you have any concerns about our products,
you can contact us on
ProductSafety@springernature.com

In case Publisher is established outside the EU,
the EU authorized representative is:
Springer Nature Customer Service Center GmbH
Europaplatz 3, 69115 Heidelberg, Germany

Printed by Libri Plureos GmbH
in Hamburg, Germany